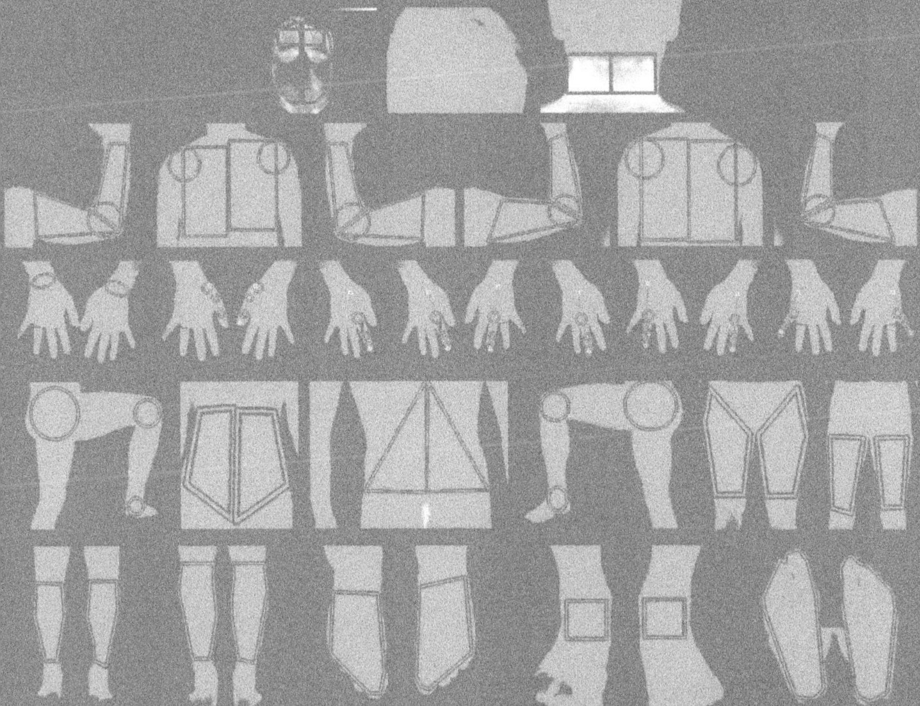

The Thermal Human Body

The Thermal Human Body

A Practical Guide to Thermal Imaging

Kurt Ammer | Francis Ring

JENNY STANFORD
PUBLISHING

Published by

Jenny Stanford Publishing Pte. Ltd.
Level 34, Centennial Tower
3 Temasek Avenue
Singapore 039190

Email: editorial@jennystanford.com
Web: www.jennystanford.com

British Library Cataloguing-in-Publication Data
A catalogue record for this book is available from the British Library.

The Thermal Human Body: A Practical Guide to Thermal Imaging

ISBN 978-981-4745-82-6 (Hardcover)
ISBN 978-0-429-01998-2 (eBook)

Contents

Preface

The book *Human Body Temperature*, by Y. Houdas and E. F. J. Ring, was a landmark in the early phase of science-based thermology. It provided a summary on the physics and physiological mechanisms involved in generation of the temperature scenario of the human body surface. The book has served as guideline for the many instructional courses on medical thermography organized in cooperation between European Association of Thermology and the University of South Wales, formerly named University of Glamorgan.

Since the book was published in 1982, we had long considered the idea to update the book and combine it with the syllabus of our thermography courses. It would also include the standardization protocol for the capture and analysis of medical infrared thermal images. A plan for the contents of the book was finally developed during the 18th Congress of the Polish Association, held in Zakopane, Poland.

However, the realization of this plan proved to be more difficult. The chapters on history, infrared thermal imagers, and quality assurance were straightforward. Also, the sections on standard protocols, provocation tests and the guidelines for recording and analysis of medical thermal images were also readily established. The topic of radiation physics proved to be more complex, since our intention was to keep to simple language and reduce formulae and laws of physics to the minimum. Since temperature and heat are often confused, an important aim of the physics chapter was to provide clear definitions of temperature and thermal energy and their relationship.

The most complex chapter was that on thermal physiology. Although the corresponding chapter in Houdas' book was useful as a guideline, there were many historical concepts; and the anatomy and physiology of neural circuits, intracellular pathways, involved transmitters and molecular processes that were unknown in the 1980s. The intention was to show that the human thermoregulation is an efficient, but very complex system that keeps deep body temperature within a narrow range. Skin temperature, which can elegantly be mapped with an infrared camera, is the most important interface between deep body temperature and the thermal ambience. It proved to be difficult to find the right balance in describing the main mechanisms of thermoregulation, and the complex processes that are involved on the cellular and sub cellular level.

Since the flow of thermal energy is not restricted to a specific body organ or structure, heat distribution throughout the body is difficult to understand. It is true that the vasculature is much involved in thermoregulatory responses, but the priority of perfusion is blood supply for nutritional demands, and thermoregulatory responses comes second.

It is now accepted, that thermal imagers should be used as measurement tools for temperature detection. Important sources of uncertainty of temperature readings from thermograms are not only the equipment itself, but also the conditions for image recording. Furthermore, the subject, their preparation for and posture during imaging, and the method of extracting temperature values from the image, all need to be accurate and reproducible. Apart from physics and physiology, all other chapters in this book propose techniques to reduce the uncertainty of temperature measurements based on thermal images.

This book provides a guide for the constantly growing community of users of medical thermal imaging. It aims to guide where and how to use infrared equipment in a strictly standardized way and ultimately to be able to comprehensively report their findings. Due to the insight in the complex mechanisms behind the distribution of surface temperature, future users of medical thermal imaging should be able to provide careful, and cautious, interpretations of infrared thermograms, thus avoiding the pitfalls of the past.

The authors wish to acknowledge the assistance of Prof. Stephen Lillicrap, medical physicist, for reviewing the physics chapter, and we thank Stanford Chong for his patience during the preparation of the manuscript.

Kurt Ammer
Vienna, Austria

Francis Ring
Bradford on Avon, UK

Chapter 1

History of Thermal Imaging from 1960

The identification of infrared radiation was made in 1800; yet, it was 78 years after the detection of "dark heat" that the first trial was conducted to investigate the amount of infrared radiation from the human skin. The first radiometry-based skin temperatures were calculated in 1921 and directly read from the radiometer in 1923 [1, 2]. Subsequently, infrared imaging was developed as a military tool, which was not declassified for civilian use until the late 1950s.

1.1 Development of Imaging Systems

In Germany, Marianus Czerny, the professor of physics at Goethe Institute Frankfurt University, laid foundations for the development of thermal sensors. One of his students in the USA, Bowling Barnes, built the first thermal imager based on thermistors in the 1950s [3]. In the UK, a research physicist Dr Max Cade had built a scanning infrared thermograph using an electronic detector of indium antimonide [4]. The prototype of this scanner was used for clinical research in the UK in 1959. Though the images were primitive, it was evident that they could show variations in skin temperature. Subsequently a more usable, but bulky device was built that produced a better quality image of the human body. However, a single hand thermogram took 5 minutes to record. Camera systems in the 1950s and 1960s made use of reflective optics with motorized mirrors to

The Thermal Human Body: A Practical Guide to Thermal Imaging
Kurt Ammer and Francis Ring

ISBN 978-981-4745-82-6 (Hardcover), 978-0-429-01998-2 (eBook)
www.jennystanford.com

perform a scan for image generation. High-speed motors produced high frequency noise and the acoustic shielding of the system was inadequate. A comparison of the systems available at that time is shown in Table 1.1. [5].

The first introduction of a commercial camera system with a lens was in the late 1960s [6]. This improvement not only reduced noise but facilitated the clinical camera, but temperature measurement remained difficult. The estimation of temperature became possible through the use of isotherms, and in 1972, the research group in Bath UK, developed a system called the thermographic Index based on measuring the area of isotherms within a specified anatomical region [7]. Some manufacturers introduced a cross-wire facility to indicate a spot temperature within the displayed image (circa 1985). At this time, personal computers became more commonly available, and also facilities to define regions of interest and employ image processing techniques. Errors introduced by analogue to digital conversion were not recognized at that time, before full digital technology became standard.

In the 1960s, indium antimonide was used as a single element detector, cooled by liquid nitrogen, that was added manually to the detector flask at regular intervals. Leidenfrost transfer systems, electronic and gas cooling were developed, which reduced the need for regular topping up of nitrogen. It is now known that some of the variables encountered with the thermal imagers could well have been due to inconsistent levels of the coolant on the detector cell. While cooled detector systems still give high performance thermal imaging, uncooled cameras have now reached a high level of thermal resolution and are more convenient in clinical applications.

1.2 Development of Applications

While the first developments were focused on medical studies, veterinary applications were also conducted, which broadened into wider biological fields.

1.2.1 Medical Applications

The first conference on medical thermography was held in New York in 1963 and another in Strasbourg, France, in 1966. Infrared imaging

Table 1.1 Thermographic Systems available in the late 1960s

System	Bofors	Aga	EMI	Rank	BarnesM1B	BarnesT6D	Barnes T4
Detector	Indium-antimonide	Indium-antimonide	Indium-antimonide	Indium-antimonide	Thermistor	Pyroelectric	Thermistor
Cooling	75 cc N_2	75 cc N_2	Leidenfrost transfer 1 L N_2	Leidenfrost transfer 5 L N_2	Not required	Not required	Not required
Scanning time	0.5 s	Instant	0.5 s 2.0 s	Instant	90 s	30 s	60 s
Field of view	25° × 12.5°	25° × 12.5° *or* 10° × 10° *or* 40° × 40°	21° × 8°	28° × 34°	10° × 15° *or* 10° × 20°	10° × 10°	20° × 10°
Focal length	0.25 m–∞	0.95 m–∞	1 m–∞	0.25 m–∞	3 ft–∞	2.5 ft–∞	60 cm 180 cm
Picture lines Elements	160 lines 40,000 elements	210 lines/frame	100 lines 25 lines	120 lines	90 lines/frame 250 elements/line	10,000 elements	60 lines 120 elements
Display size	68 × 34 mm *or* 68 × 68 mm	90× 90 mm *or* 67 × 74 mm	100 × 60 mm	145 × 114 mm	Polaroid	95 × 72 mm Polaroid only	Polaroid

(*Continued*)

Table 1.1 (*Continued*)

System	Bofors	Aga	EMI	Rank	BarnesM1B	BarnesT6D	Barnes T4
Frame rate	2/s	16/s	2/s 0.5/s	46/s	single frames	single frames at 30 s	single frames
Camera	35 mm Polaroid	35 mm Polaroid	35 mm Polaroid	35 mm	Polaroid	Polaroid	Polaroid
Recording	Photographic intensity modulation A-scan	Photographic intensity modulation	Photographic intensity modulation A-scan	Photographic intensity modulation, separate A-scan	Polaroid film only	Direct to colour Polaroid film	Direct to Polaroid film
Isotherm	Four widths 2.5–20% of range set	Single adjustable				Minimum range 2.5°C integral colour range	Eight step isotherms extra
Line-scan	Output from display unit for oscilloscope		Integral with system display	Separate recorder	Analogue signal for recorder Thermoscope	Vertical scan locked for oscilloscope or chart recorder	Photocomparator for thermograms, extra

in medicine was in its infancy, but those early proceedings showed the variety and detail, given the relatively primitive state of the technology. One author concluded, "All that has been revealed by this technique is nothing compared with what is left to be discovered" [8]. More than 60 years have passed since this was written, and certainly more has been revealed, but we may well repeat the optimistic statement from 1963 today!

As the proceedings and publications of the early years (1960–1970s) show the predominant interest was in the possible screening and early diagnosis for breast cancer by thermography [9–11]. This interest gradually reduced with the arrival of alternative breast imaging modalities. Skin malignancies, e.g. melanoma, have been another application of diagnostic thermography from the beginning [12] to the present day [13]. Diagnosis of Raynaud's phenomenon and the response to a cold challenge have been successful applications of thermography. The need for standardization was realized from the outset [14, 15]. Other vascular conditions included venous disorders [16] and tissue viability [17, 18]. Studies in plastic surgery have recently re-confirmed the value of thermal imaging for tissue viability assessment [19]. Infrared-thermography and other techniques of mapping temperature distribution on surface of the body has been applied in spinal disorders and related neurological syndromes for more than 40 years. However, the validity of thermal imaging for diagnosing back disorders is still an ongoing debate [20]. Some investigators have tried to find a relationship between skin temperature and innervation fields of sensory nerve fibres, but this has not been established.

In recent years following pandemic fever outbreaks, e.g. severe acute respiratory syndrome (SARS) and avian flu, the need for fever screening in international travel led to the employment of thermal imaging. The International Organization for Standardization was required to address the issue of minimum standards needed to detect the presence of fever [21].

Thermal imaging has been first used as an outcome measurement in the field of rheumatology. Different dosage of anti-inflammatory compound drugs was objectively compared in clinical trials. At that time, little attention had been given to standardization of procedures. However, standardized methodology was required to meet the essentials of good clinical practice [22, 23]. In addition to anti-inflammatory drugs, a number of thermographic studies have shown the effect of vasoactive drugs on skin temperature [24–26].

1.2.2 Applications in Biology

Thermography in veterinary medicine was included in the first international symposium in New York 1963 [27]. Other early applications included animal physiology [28, 29], animal models of physiological functions [30] and pharmacology [31]. Further diagnostic thermography was developed for domestic and farm animals [32, 33]. Specialized centres around the world have extended thermal imaging studies in animals, i.e. cattle, horses, pigs, etc.

Historical bibliographies are useful resource to understand the past and extent those results with current technology. The following are two particularly relevant sources:

- *International Bibliography of Medical Thermology*, by Margaret Abernathy and Thomas Bradford Abernathy, *Thermology*, 1987, vol. 2, no. 3, pp. 117–534. ISSN; 0882-3758.
- DVD: *Publications on Thermology 1989 to 2012*, email contact: Kammer1950@aol.com.

References

1. Aldrich LB. The Melikeron: An approximately black-body pyranometer. *Smithsonian Miscellaneous Collections* 1922, 72(13)3–11.
2. Aldrich LB. A study of body radiation. *Smithsonian Miscellaneous Collections* 1928, 81(6), 1–54.
3. Barnes RB. Thermography, Thermography & its clinical applications. *Annals of the New York Academy of Sciences* 1964, 121, 34–48.
4. Cade CM. High-speed thermography. Thermography & its clinical applications. *Annals of the New York Academy of Sciences* 1964, 121, 71–79.
5. Ring EFJ. Equipment for medical thermography. *British Journal of Hospital Medicine* 1971, 5, equipment suppl, 24.
6. Dibley D. Opto-mechanical systems for IR thermography. In: Ammer K, Ring EFJ, eds, *The Thermal Image in Medicine and Biology*. Uhlen Verlag Wien, 1995, pp. 33–39.

7. Winsor T. Bendbezu J. Thermography & the peripheral circulation. Thermography & its clinical applications. *Annals of the New York Academy of Sciences* 1964, 121, 135–156.
8. Collins AJ, Ring EFJ, Cosh JA, Bacon PA. Quantitation of thermography in arthritis using multi isothermal analyis. 1. The thermographic index. *Annals of Rheumatic Diseases* 1974, 33, 113–115.
9. Williams KL, Williams FL, Handley RS. Infra-red thermometry in the diagnosis of breast disease. *The Lancet* 1961, 278(7217), 1378–1381.
10. Melander OLOF. Thermography as the primary step in mass screening for breast cancer. *Bibliotheca Radiologica* 1975, (6), 91–96.
11. Stark AM et al. Malignant diseases section, in Ring EFJ, Phillips B, eds, *Recent Advances in Medical Thermology*, Plenum Press, New York, London, 1984, pp. 549–634.
12. Lintermans J, Maillard GF. Essai d'explication thermochimique de l'hyperthermie. *Archiv für dermatologische Forschung* 1971, 240, 237–244.
13. Di Carlo A, Elia F, Desiderio F, Catricalà C, Solivetti FM, Laino L.. Can video thermography improve differential diagnosis and therapy between basal cell carcinoma and actinic keratosis? *Dermatologic Therapy* 2014, 27(5), 290–297.
14. Ring EFJ, Aarts NJM, Black CM, Bösiger P, et al. Raynaud's Phenomenon: Assessment by thermography (EAT-report). *Thermology* 1988, 3, 69–73.
15. Ammer K. Cold challenge to provoke a vasospastic reaction in fingers determined by temperature measurements; a systematic review. *Thermology International* 2009, 19(4) 109–118.
16. Patil KD, Williams JR, Williams, KL. Thermographic localization of incompetent perforating veins in the leg. *BMJ* 1970, 1(5690), 195–197.
17. Henderson HP, Hackett, ME. The value of thermograph in peripheral vascular disease. *Angiology* 1978, 29, 65–75.
18. Spence VA, McCollum PT, Walker WF, Murdoch G. Assessment of tissue viability in relation to the selection of amputation level. *Prosthetics and Orthotics International* 1984, 8(2), 67–75.
19. de Weerd L, Mercer JB, Weum S. Dynamic infrared thermography. *Clinics in Plastic Surgery* 2011, 38(2), 277–292.
20. Ammer K. Thermography in spinal disorders: A narrative review. *Thermology International* 2010, 20(4), 117–125.

21. Ring EFJ, McEvoy H, Jung A, Zuber J, Machin G. New standards for devices used for the measurement of human body temperature. *Journal of Medical Engineering & Technology* 2010, 34(4), 249–253.

22. Bacon PA, Ring EFJ, Collins AJ. Thermography in the assessment of anti rheumatic agents. In: Gordon JL, Hazleman BL, eds, *Rheumatoid Arthritis*. Amsterdam: Elsevier/North Holland Biomedical Press, 1977, pp. 105–110.

23. Bird HA, Ring EFJ, Bacon PA. A thermographic and clinical comparison of three intra-articular steroid preparations in rheumatoid arthritis. *Annals of Rheumatic Diseases* 1979, 38, 36–39.

24. Ring EFJ, Owens CWI, Cosh JA; Eghtedari A. Infrared techniques in the assessment of peripheral temperature response to anti-hypertensive drugs. In: Birley DM, ed. *The Clinical Impact of Beta-Adreno-Receptor Blockage*. Horsham, UK, 1980, pp. 267–273.

25. Ring EFJ, Porto LO, Bacon PA. Quantitative thermal imaging to assess inositol-nicotinate treatment for Raynaud's syndrome. *Journal of International Medical Research* 1981, 9, 393–400.

26. Giacovazzo M, Valducci G, Martelletti P, Romiti A, Gallo MF, Valducci M. Telethermographic investigations in migraine patients before and after flunarizine treatment. *Cephalalgia* 1985, 5 (2 suppl), 169–171.

27. Smith WM. Applications of thermography in veterinary medicine. *Annals of the New York Academy of Sciences*, 1964, 121(1), 248–254.

28. Cena K, Clark JA. Thermographic measurements of the surface temperature of animals. *Journal of Mammalogy* 1973, 54, 1003–1007.

29. Szekely M, Kellermayer M, Donhoffer S. Thermoregulatory role of periaortic brown adipose tissue in rats acclimated to different ambient temperatures. *Acta Physiologica Academiae Scientiarum Hungaricae* 1970, 40 (3), 251–260.

30. Collins AJ, Ring EFJ. Measurement of inflammation in man and animals by radiometry. *British Journal of Pharmacology* 1972, 44(1), 145–152.

31. Malm A, Arborelius M, Bornmyr S, Lilja B, Gill RL. Effects of oxygen on acute myocardial infarction: A thermographic study in the dog. *Cardiovascular Research*, 1977, 11(6), 512–518.

32. Stromberg B. Thermography in veterinary medicine. *Bibliotheca Radiologica* 1975, 6, 231–236.

33. Purohit RC. History and research review of thermology in veterinary medicine at Auburn University. *Thermology International* 2007, 17(4) 127–132.

Chapter 2

Basic Thermal Physics: Heat Exchange and Infrared Radiation

This chapter is dedicated to basic thermal physics, including definitions of temperature and heat energy. The Laws of Thermodynamics define energetic relationships within thermodynamically closed (no material exchange) and isolated structures (no material or energy exchange). Although living beings are rarely (if ever) in those states, these laws still apply, and provide the scientific foundation for understanding temperature measurement. They also define the principles of heat transfer. The description of the principles of heat transfer follows closely Chapter 2 of the book *Human Body Temperature*, by Y Houdas and EFJ Ring.

2.1 What Is Temperature?

Thermodynamic temperature is one of seven basic quantities which together with length, mass, time, electric current, amount of substance and light intensity is used to generate the basics of the measurement and dimension system of physics [1, 2]. Units of physics, necessary to understand the mechanisms involved in heat transfer are listed in Table 2.1. This list is based on the SI Units [2]. The base unit temperature is an *intensive* quantity of heat energy. A physical property of a system that does not depend on the system size or the amount of material in the system is called intensive. Although temperature represents the mean kinetic energy of atoms or molecules in a substance, temperature is not the same as heat.

The Thermal Human Body: A Practical Guide to Thermal Imaging
Kurt Ammer and Francis Ring

ISBN 978-981-4745-82-6 (Hardcover), 978-0-429-01998-2 (eBook)
www.jennystanford.com

Table 2.1 Système Internationale (SI): basic units [2]

Quantity	Definition	Symbol for quantity	SI units	Abbreviations
Length	The meter is the length of the path travelled by light in vacuum during a time interval of 1/299 792 458 of a second.	*l*, *x*, *r*, etc.	meter	m
Mass	The kilogram is the unit of mass; it is equal to the mass of the international prototype of the kilogram.	m	kilogram	kg
Time	The second is the duration of 9 192 631 770 periods of the radiation corresponding to the transition between the two hyperfine levels of the ground state of the caesium 133 atom.	t	second	s
Electric Current	The ampere is that constant current which, if maintained in two straight parallel conductors of infinite length, of negligible circular cross section, and placed 1 meter apart in vacuum, would produce between these conductors a force equal to 2 × 10−7 newton per meter of length.	I, i	ampere	A
Thermodynamic temperature	The kelvin, unit of thermodynamic temperature, is the fraction 1/273.16 of the thermodynamic temperature of the triple point of water.	T	kelvin	K
Amount of Substance	1. The mole is the amount of substance of a system which contains as many elementary entities as there are atoms in 0.012 kilogram of carbon 12; its symbol is "mol." 2. When the mole is used, the elementary entities must be specified and maybe atoms, molecules, ions, electrons, other particles, or specified groups of such particles.	n	mole	mol
Luminous Intensity	The candela is the luminous intensity, in a given direction, of a source that emits monochromatic radiation of frequency 540 × 1012 hertz and that has a radiant intensity in that direction of 1/683 watt per steradian.	I_v	candela	cd

Table 2.2 Systeme Internationale (SI): coherent derived Units [2]

Derived quantity	Name	Symbol	Expressed in terms of other SI units	Expressed SI in terms of base units
Plane angle	Radian	rad		m/m
Solid angle	Steradian	sr		m^2/m^2
Area	Square meter	A		m^2
Volume	Cubic meter	V		m^3
Speed, velocity	Meter per second	v		m/s
Angular velocity	Radian per second	rad/s		$m\ m^{-1}\ s^{-1} = s^{-1}$
Acceleration	Meter per second squared	a		m/s^2
Angular acceleration	Radian per second squared	rad/s^2		$m\ m^{-1}\ s^{-2} = s^{-2}$
Wavenumber	Reciprocal meter	$\sigma, \tilde{\nu}\ \nu$		m^{-1}
Density, mass density	Kilogram per cubic meter	r		kg/m^3
Surfacc density	Kilogram per square meter	ρ_A		kg/m^2
Specific volume	Cubic meter per kilogram	v		m^3/kg
Mass concentration	kilogram per cubic meter	ρ, γ		kg/m^3
Pressure, stress	pascal	Pa	N/m^2	$m^{-1}\ kg\ s^{-2}$
Dynamic viscosity	pascal second	Pa s		$m^{-1}kg\ s^{-1}$
Force	newton	N		$m\ kg\ s^{-2}$
Moment of force	newton meter	Nm		$m^2\ kg\ s^{-2}$

(*Continued*)

Table 2.2 (*Continued*)

Derived quantity	Name	Symbol	Expressed in terms of other SI units	Expressed SI in terms of base units
Surface tension	newton per meter	N/m		kg s^{-2}
Frequency	hertz	Hz		s^{-1}
Celsius temperature	degree Celsius	°C	0°C = 273.16 K	K
Energy, work, amount of heat	joule	J	N m	kg m^2 s^{-2}
heat capacity, entropy	joule per kelvin	J/K		m^2 kg s^{-2} K^{-1}
specific heat capacity, specific entropy	joule per kilogram kelvin	J/(kg K)		
specific energy	joule per kilogram	J/kg		m^2 s^{-2}
energy density	joule per cubic meter	J/m^3		m^{-1} kg s^{-2}
molar energy	joule per mole	J/mol		m^2 kg s^{-2} mol^{-1}
molar entropy, molar heat capacity	joule per mole kelvin	J/(mol K)		m^2kg s^{-2} K^{-1} mol^{-1}
power, radiant heat, heat flux	watt	W	J/s	kg m^2 s^{-3}
heat flux density, irradiance	watt per square meter	W/m^2		kg s^{-3}
thermal conductivity	watt per meter kelvin	W/(m K)		m kg s^{-3} K^{-1}

Derived quantity	Name	Symbol	Expressed in terms of other SI units	Expressed SI in terms of base units
radiant intensity	watt per steradian	W/sr		$m^4\ m^{-2}\ kg\ s^{-3} = m\ kg\ s^{-3}$
radiance	watt per square meter steradian	$W/(m^2\ sr)$		$m^2\ m^{-2}\ kg\ s^{-3} = kg\ s^{-3}$
luminance	candela per square meter	*Lv*		cd/m^2
luminous flux	lumen	lm	cd sr	Cd
illuminance	lux	lx	lm/m^2	m^{-2} cd

Table 2.3 SI prefixes [2]

Factor	Name	Symbol	Factor	Name	Symbol
10^1	deka	da	10^{-1}	deci	d
10^2	hecto	h	10^{-2}	centi	c
10^3	kilo	k	10^{-3}	milli	m
10^6	mega	M	10^{-6}	micro	μ
10^9	giga	G	10^{-9}	nano	n
10^{12}	tera	T	10^{-12}	pico	p
10^{15}	peta	P	10^{-15}	femto	f
10^{18}	exa	E	10^{-18}	atto	a
10^{21}	zetta	Z	10^{-21}	zepto	z
10^{24}	yotta	Y	10^{-24}	yocto	y

2.1.1 Thermodynamic Temperature

The unit of thermodynamic temperature is the kelvin. One kelvin is the fraction 1/273.16 of the thermodynamic temperature of the triple point of water. This definition was supplemented in 2005 by the description of the isotopic composition of water, exactly defined by the amount of substance ratios. The *triple point* of a substance is described by the temperature and pressure at which the three phases (gas, liquid, and solid) of that substance coexist in thermodynamic equilibrium.

2.1.2 Celsius Temperature

It remains common practice to express a thermodynamic temperature, symbol T, in terms of its difference from the reference temperature T_0 = 273.15 K, the ice point of water. This difference is called the Celsius temperature, symbol t, which is defined by the quantity equation:

$$T = T - T_0.$$

The unit of Celsius temperature is the degree Celsius, symbol °C, which is by definition equal in magnitude to the kelvin. A difference or interval of temperature may be expressed in kelvin or in degree Celsius the numerical value of the temperature difference being the same.

2.2 What Is Energy?

Energy can be defined as the ability to do work and is expressed in the same equation as work. It exists in many forms, such as mechanical, electrical, nuclear, and thermal [3].

Transformation of one form of energy into another is possible, for example, a thermally powered generating station can produce electricity. Such transformation is never totally efficient. However, transformation of all other forms of energy into heat can be total, but the contrary, i.e., the transformation of heat into one of the other forms, is never 100%. Heat thus appears to be a particular form of energy.

The SI unit of energy is the Joule. The unit of different forms of energy is separately defined.

For *mechanical energy*, 1 joule is equal to the work done when a force of one newton moves the point of its application a distance of one metre in the direction of the force.

$$1\ \text{J} = 1\ \text{m}\cdot\text{N},$$

where N is the force capable of giving a mass of 1 kg an acceleration of 1 meter per second, per second.

The unit Joule for *electrostatic potential energy* is expressed by

$$1\ \mathrm{J} = U_E = C\ V^2/2,$$

where Coulomb (C) is the amount of electricity carried in one second of time by 1 ampere of current and voltage (V) is the difference in electric potential across two points along a conducting wire carrying 1 ampere of constant current when the power dissipated between the points equals 1 watt.

The *thermoelectric effect* encompasses three separately identified effects: Seebeck effect, Peltier effect, and Thomson effect. These effects describe the direct conversion of temperature differences to electric voltage and vice versa.

For *thermal energy*, 1 Joule is expressed as the 1/4.184 part of heat energy required to raise the temperature of a unit weight (1 g) of water from 0°C to 1°C. 4.184 Joules are equal to the traditional unit of 1 calorie.

There are relationships between heat and electromagnetic radiation, and thermal energy and chemical energy.

Above the absolute zero-point of –273.15°C, everything, solid material, fluids, gases and living beings, emits to some extent *radiant heat* in the form of infrared waves. *Chemical energy* is related to the structural arrangement of atoms or molecules and represents the energy of chemical bonds and is equal to the amount of the heat required to break one mole of molecules into their individual atoms. *Potential energy* is a property of a system representing stored energy that depends upon the relative position of various parts of a system.

2.3 What Is Power?

Physically, power [W] is work or *energy per time unit*. In the case of mechanical energy per time unit it is called mechanical power or, particularly in thermal, sports and occupational physiology, *work rate* and attributed the letter *W* [4]. The heat equivalent to power is *heat flux* with the official unit *watt* (W), which is the power corresponding to 1 joule per second. The kilocalorie per hour (kcal/h) must be avoided.

Understanding power enables to follow the process of heat exchange.

2.4 Energy Conversion and Heat Transfer

An excellent description of the history of thermodynamics was published by I. Müller [5].

2.4.1 First Law of Thermodynamics

The *first law of thermodynamics* states that the various forms of energy can be changed· from one form to another but cannot be created or destroyed. In other words, the total sum of the energy is stable over time irrespective of number of energy transformations. If a system has received, say, mechanical energy, *W*, and if its internal energy, *E*, has not changed, the system must has lost an equivalent amount of some other energy, e.g. heat, *Q*. There is always a balance between the energy gained, which is counted positively, and the energy lost, which is scored negatively:

$$W = -Q$$

When the internal energy of the system has varied from the value E_1 to the value E_2, the following equation is derived:

$$E_2 - E_1 = W + Q$$

A practical application in medicine is the fact that weight gain is secondary to increased calorie intake and/or decreased energy expenditure [6].

2.4.1.1 Energy release in living tissues

The energy possessed and exchanged by living creatures is made up from dynamic(kinetic) and static forms (potential: mass-, chemical-, and force-related energies). Energy release in living tissues is based on the chemical oxidation-reduction reaction. The term *oxidation* conveys the impression of a reaction involving combination with oxygen:

$$(CH_2O)_n + nO_2 \rightarrow nCO_2 + nH_2O + \text{Energy}$$

In this example, the molecules of carbohydrates are oxidized by the addition of oxygen atoms. However, oxidation is more generally defined as the removal of electrons from a molecule. In many reactions, the electrons are provided in the form of hydrogen atoms, and the oxidized compound is thereby dehydrogenated.

Conversely, *reduction* is the addition of hydrogen atoms to a molecule or more generally the gain of electrons, as in the formation of ammonia:

$$N_2 + 3H_2 \rightarrow 2NH_3$$

In biological reactions, a reduction occurs for every oxidation reaction because the hydrogen atoms released by oxidation do not remain free. They are immediately tied up by other atoms or molecules. Thus, the general concept of energy release can be represented as follows [3]:

AH_2	+ B	→ A	+ BH_2	+ Energy
Substance to be oxidized	Substance to be reduced	Substance oxidized	Substance reduced	To be captured and/or transformed
(potential energy)	(generally O_2)	(generally CO_2)		

2.4.1.2 Enthalpy

Internal energy is the property or state function that defines the energy of a substance in the absence of effects due to external electric, magnetic, and other fields. Different to temperature, internal energy is an extensive property- that is, its magnitude depends on the amount of substance in a given state. Sometimes it is convenient to represent the internal energy as a sum of terms that can be interpreted as kinetic energy, potential energy, and chemical energy.

The sum of the internal energy and the product of the pressure and volume of a thermodynamic system is called *enthalpy*. It is an energy-like property or state function with the dimension of energy. Houdas & Ring provided the following explanation of enthalpy [3].

Consider a system in which a chemical reaction occurs. The initial internal energy is E_1 and the final internal energy is E_2. If the reaction is carried out at constant volume (no gas expansion), there is no mechanical work. Therefore,

$$E_2 - E_1 = Q_v,$$

where the subscript v indicates that the reaction occurs at constant volume. If Q_v is positive, the reaction is said to be *endothermal*. If Q_v

is negative, the reaction is *exothermal.* The combustion of food in a bomb calorimeter is the classic example of an exothermal reaction at constant volume in a thermally insulated system (*adiabatic reaction*). In living bodies, reactions generally occur at constant pressure. It is thus possible to write

$$E_2 - E_1 = W + Q_p.$$

For a gas, the mechanical energy W corresponds to the product of pressure and volume, $P \cdot V$.

Thus,

$$W = -P\,(V_2 - V_1),$$

which then becomes

$$E_2 - E_1 = PV_1 - PV_2 + Q_p$$

or

$$Q_p = (E_2 + PV_2) - (E_1 + PV_1).$$

The term $E + PV$ *is* called the *enthalpy* and is symbolized by H. Enthalpy has the dimension of energy. Therefore,

$$Q_p = H + W.$$

For example, the complete oxidation of 1 mole of D-glucose at 25°C and 760 mm Hg produces 673 kcal (2813 kJ) according to the following reaction:

$$C_6H_{12}O_6 + 6O_2 \rightarrow 6CO_2 + 6H_2O - 2813\ \text{kJ}$$

Note that the negative sign is used before 2813 kJ (= 673 kcal) because heat is given off by the system (exothermal reaction).

In this reaction, there is no change in gas volume. Six litres of O_2 is used and 6 litres of CO_2 is produced. According to the law of energy conservation, the change in internal energy is equal to the heat transferred to, less the work done by, the system. When no work is done to change the gas volume, $W = 0$, the change of internal energy, ΔE, is equal to the systems heat content at constant volume, Q_v and therefore is the same as the change of enthalpy, ΔH.

$$\Delta E = Q_v = \Delta H$$

For these kinds of chemical compounds, ΔH can be measured by the combustion of the substance in a calorimetric bomb. Since the energy given off by this combustion is the same whatever the route of oxidation, ΔH is the same in the calorimetric bomb as in the living body. Tables exist giving the values of ΔH for the most common compounds that can be oxidized in the body.

In other words, the total amount of energy possessed by an object is known as its enthalpy, which is minimal (but not absent) at temperatures approaching absolute zero. It varies with the pressure and volume (mass) of each system, and its kinetic component causes sub-atomic and cellular movement and collisions, releasing thermal energy. Thus, heat content is a function of this collision frequency and it is quantified using temperature measurements and calorimetry [7].

2.4.2 Second Law of Thermodynamics

When a warm object is put into cold water, its temperature falls, but the water temperature increases, until the two become equal. This is due to energy transfer from the high level (high temperature) to the low level (low temperature) [3]. However, the reverse does not occur spontaneously. An analogy can be drawn with the mechanical energy produced in a ball rolling down a staircase. The ball goes down from one level to a lower one because gravitational energy is higher at the upper level. In this case, the reverse transfer of energy cannot occur spontaneously

The same process occurs in any closed (*inanimate*) system with an outer membrane (*diathermal wall*) permissive to energetic, but not to material exchange. If that system was placed within a stable environment, the collision frequency of its particles would eventually stabilize, and a state of thermal equilibrium (steady state) would exist. The temperature of that object would now be constant, whilst particle motion continued. If another system with a lower enthalpy comes into physical contact with the former, energy will be exchanged across their contacting walls towards the latter. That is, thermal energy moves down energy gradients, either through a change of state (solid, liquid, gas) or via conductive (molecule to molecule), convective (mass flow) or radiative transfers. This establishes thermal gradients within and between these systems, with both systems eventually attaining a common thermal equilibrium.

The first law of thermodynamics expresses the equivalence between the various forms of energy, but it cannot predict the spontaneous evolution of a reaction or an energy exchange. This prediction is the purpose of the *second law of thermodynamics,* first proposed in 1824 by Carnot. In its present form, this law states that energy transformation can occur spontaneously only in one direction, that is, toward greater entropy [3].

Entropy is the measure of a system's thermal energy per unit temperature that is unavailable for doing useful work. Because work is obtained from ordered molecular motion, the amount of entropy is also a measure of the molecular disorder, or randomness, of a system. The idea of entropy provides a mathematical way to encode the intuitive notion of which processes are impossible, even though they would not violate the fundamental law of conservation of energy.

For example, a block of ice placed on a hot stove surely melts, while the stove grows cooler. Such a process is called irreversible because no slight change will cause the melted water to turn back into ice while the stove grows hotter. In contrast, a block of ice placed in an ice-water bath will either thaw a little more or freeze a little more, depending on whether a small amount of heat is added to or subtracted from the system. Such a process is *reversible* because only an infinitesimal amount of heat is needed to change its direction from progressive freezing to progressive thawing. Similarly, compressed gas confined in a cylinder could either expand freely into the atmosphere if a valve were opened (an irreversible process), or it could do useful work by pushing a moveable piston against the force needed to confine the gas. The latter process is reversible because only a slight increase in the restraining force could reverse the direction of the process from expansion to compression. For reversible processes, the system is in *equilibrium* with its environment, while for irreversible processes it is not.

To provide a quantitative measure for the direction of spontaneous change, the German physicist Clausius introduced the concept of entropy as a precise way of expressing the *second law of thermodynamics.*

If we assume a reversible energy transformation, in which the system exchanges thermal energy Q with its environment at temperature T, the entropy of the system varies from S_1 to S_2 and this variation can be written

$$S_2 - S_1 = \Delta S = Q_{\text{rev}}/T$$

(The subscript "rev" means that the reaction is reversible.) This case is ideal. In practice, the reactions are not reversible; in this case ΔS is greater than Q_{irr} (the subscript "irr" means that the reaction is irreversible).

If we now consider an insulated system in which a reversible reaction has occurred, entropy has not varied because $Q_{rev} = 0$. Thus we have $\Delta S = 0$. If the transformation occurs spontaneously, we now have $Q_{irr} = 0$. This necessarily implies that ΔS is positive. This conclusion is very important: *The only possibility for spontaneous variation of the entropy of a closed insulated system is its increase* [3].

If we consider again the example of a warm object put into cold water, we see that the maximum entropy will correspond to thermal equilibrium. The same reasoning can be applied to all other energy transformations. Total equilibrium always corresponds to the maximal entropy. Therefore, the gradual evolution of a living body into physical, chemical, and thermal equilibrium with its environment is death. On the contrary, life, which is the creation of new molecules with the storage of potential energy, corresponds to negative entropy. This is physically possible only because the living body is not a closed insulated system. It receives energy from the sun, either directly or indirectly [3].

2.4.2.1 Free Enthalpy, Usable Energy

When a molecule of food is oxidized, a certain amount of energy depends only on the initial state (the molecule of food) and its final state (the waste molecules), but is independent of the intermediate states. We have seen that the amount of energy produced could be considered as a variation in the enthalpy *H*. However, this energy cannot be completely transformed into mechanical, chemical, electrical, or other types of work. The major part of this energy must appear in the form of heat and corresponds to the variation of entropy [3].

Free enthalpy or *free energy* is that fraction of energy that can be transformed into work. It is symbolized by *G* (or ΔG) and is defined according to the Gibbs equation as follows:

$$\Delta G = \Delta H - T\Delta S$$

As ΔS is always positive (except in an ideal case), $T\Delta S$ is also always positive, except at absolute zero. Therefore, all living bodies produce heat, which can be considered as "physical waste". On the other hand, if $T\Delta S$ is positive, this implies that ΔG is greater than ΔH (in absolute values).

Such reaction is termed *exoergic* because it can supply the energy for work. For example, in the complete oxidation of 1 mole of n-glucose at a temperature of 298 K, ΔG is −686.48 kcal and ΔH is −673.00 kcal. Entropy has then been increased by (686.48 − 673.00)/298, i.e. 45.2 e.u. (e.u. = entropy unit, expressed in kcal/K).

On the contrary, the synthesis of 1 mole of n-glucose corresponds to a negative entropy ($T\Delta S$ = −13.480 kcal). This reaction is said to be *endoergic* and can occur in living bodies only because an equivalent amount of energy has been supplied by the environment.

In the thermal function of homeotherms, only that part of energy which appears as heat is directly involved. For this reason, the next section will show how heat is exchanged within the body and / or between the body and its environment.

2.4.3 Ways of Heat Transfer

The study of heat transfer is called *thermokinetics.* When there is a difference of temperature between two bodies or between parts of the same body, heat flows from hot areas to cold areas spontaneously so that the temperature difference between them is diminished toward zero. Newton's Law (1701) describes how a difference in temperature between two bodies is equalized. Heat flows from hot to cold spontaneously until the difference is zero.

Variation in $T_b = T_b/dt = \text{Kn}\,(T_b - T_a)$

where T_b represents the warm body and T_a is the cool environment. Kn is Newton's cooling constant, and dt = temperature difference.

A temperature variation thus occurs because heat has been transferred from the body to its environment. This heat transfer depends directly on the temperature difference between the body and its environment. When T_b becomes equal to T_a, there is no further heat transfer. The system is then in thermal equilibrium. If the body itself is producing heat continuously, T_b always remains higher than T_a.

If this temperature difference does not vary according to time, the system is said to be in a *steady state.* Two possibilities for heat transfer are described in physics:

1. *Heat transfer without change of state.* This can occur by either conduction, convection, or radiation.
2. *Heat transfer with change of state.* For example, to achieve conversion of water from liquid to vapour by boiling requires heat transfer, although the water may remain at a constant temperature.

Newton's Law considers mainly heat exchange by conduction and convection. Modification of Newton's Law has been made by Dulong & Petit and Josef Stefan [8]. Fourier proposed a different equation for heat conduction [9]. This equation, often confused with Newton's Law [10], was endorsed as basis for calorimetric measurements in animals [11].

2.4.3.1 Heat transfer without change of state

2.4.3.1.1 *Conduction*

In general, during any period in which temperatures change *in time* at any place within an object, the mode of thermal energy flow is termed *transient conduction* Heat transfer by *conduction* occurs within a solid or between two or more solids in close contact.

For example, if a metal bar has one end at temperature T_2 and the other at temperature T), in a steady state heat flows from the warm end (T_2) to the cold end (T)). The heat flow rate $\dot{Q}_k$ throughout a transverse section of area A_k is described by the relationship proposed by Fourier in 1822:

$$\dot{Q}_k = -(k/x)(T_2-T_1)A_k,$$

where k is the thermal conductivity of the material and x is the distance between the ends of the bar.

In thermal physiology, the ratio k/x is called *conductive heat transfer coefficient*, h_k, defined as the net heat transfer by conduction per unit area between a surface and a solid or stationary fluidic medium in contact with the surface per unit temperature difference (ΔT) between the surface and the substance with which it is in contact. $hk = K \cdot \Delta T^{-1}$ expressed in $W \cdot m^{-2} \cdot {}^{\circ}C^{-1}$ [4].

Its inverse ratio x/k is then given in $m^2 \cdot {}^{\circ}C \cdot W^{-1}$, and represents *thermal resistance* and can be symbolized by R_k. It is thus possible to express. heat transfer by conduction using a formula similar to Ohm's law, which is well known in electricity:

$\dot{Q}_k = (1/R_k)T \qquad I = (1/R)V$

Heat transfer Ohm's law

The concept of thermal resistance (or *impedance*) corresponds to that of *thermal insulation* and is very useful in practice because it can be measured. When heat flows through different structures or materials with different thermal insulations (for example, a clothed subject), the total thermal resistance and thus the total conduction coefficient can be calculated.

Applying the analogy with Ohm's law, the total thermal resistance of n layers having respective resistances of R_{k1}, R_{k2}, ..., R_{kn} is

$$R_k = R_{k1} + R_{k2} + \ldots + R_{kn}$$

On the other hand, in physiology, the problem of heat transfer by conduction is more complex, because (1) a steady state is rarely achieved, (2) heat is produced directly by the body, and (3) transfer occurs in three spatial dimensions.

Consider a small cube within a body. At any given point in this cube, the temperature can be related to the three coordinates, *x, y,* and *z.* The heat flowing into one face, for instance, that corresponding to the *x* axis, is described by the following equation:

$$\dot{Q}_x = -k\frac{dT}{dt}(d_y, d_z)$$

The heat flowing out from the opposite surface is $\dot{Q}_k + dx$, the difference corresponding to the heat produced by the cube and to the external variation in temperature of the cube.

If a similar reasoning is applied to all three axes, the final equation is

$$\underbrace{\frac{d^2T}{dx^2} + \frac{d^2T}{dy^2} + \frac{d^2T}{dz^2}}_{\text{Heat flowing throughout the three faces}} + \underbrace{\frac{\dot{Q}_0}{k}}_{\text{Heat produced}} - \underbrace{\frac{\rho c}{k}\cdot\frac{dT}{dt}}_{\text{Variation in temperature}} = 0$$

The inverse of the ratio $\rho c/k$ is generally termed *thermal diffusion* or in other words **thermal diffusivity** $\alpha = k/\rho cp$ ($m^2 \cdot s^{-1}$) **is** the **thermal conductivity** of a substance divided by its density ρ and **specific heat** at constant pressure c_p, [4]. Thermal diffusivity is a measure of how quickly a material can carry heat away from a hot source. Since material does not just transmit heat but must be warmed by it as well, α involves both the conductivity, k, and the volumetric heat capacity, ρc [12].

However, thermal diffusivity must be differentiated from *mass diffusivity D*, which is the constant of proportionality relating the rate of diffusion of a gas to the gradient of its concentration in another gas, e.g. water vapour in air [4]. If the rate of diffusion of mass is $\dot{m}\cdot$ [kg·s^{-1}] in a direction specified by a coordinate y [m] and if the concentration over a plane at a given value of y is p [kg·m^{-3}], the rate of flow through an area A is given by the Fickian equation,

$$\dot{m}\cdot = -DA(\partial p/\partial y),$$

where D is the **mass diffusivity** or *diffusion coefficient* and $\partial p/\partial y$ is the appropriate concentration gradient. More complex equations are needed to describe diffusion in two or three directions simultaneously [m^2·s^{-1}].

This equation for thermal diffusivity expresses the fact that the sum of the heat entering the three faces during time *dt*, added to the spontaneous heat production during the same time, is equal to the sum of heat dissipation from the three faces, plus the algebraic variation in heat content. This is of fundamental importance because all other equations can be derived from it by mathematical procedures, e.g. differentiation and integration.

For instance, if no heat is produced, the term equals $\dot{Q}_0$ zero and the equation becomes

$$\frac{d^2T}{dx^2}+\frac{d^2T}{dy^2}+\frac{d^2T}{dz^2}=\frac{1}{\alpha}\cdot\frac{dT}{dt}=0$$

This is the general expression for Fourier's law. Another example: If there is no heat source and if the system is in steady state, the repartition of temperatures within the cube follows Laplace's equation:

$$\frac{d^2T}{dx^2}+\frac{d^2T}{dy^2}+\frac{d^2T}{dz^2}=0$$

In practice, the equations given in this chapter have been used particularly to determine the thermal characteristics and thermal production of tissues, particularly when pathological, such as a tumour [13].

In modern notation, especially in the field of thermal engineering Fourier's law is written as

$$\nabla \cdot \mathrm{K}\nabla \mathrm{T} = c\frac{\partial T}{\partial t},$$

where K is thermal conductivity, T is temperature, c is specific heat capacity of the solid per unit volume, and t is time. The dependent variable T is a scalar potential, while thermal conductivity and specific heat capacity are empirical parameters [11].

2.4.3.1.2 *Convection*

Heat transfer by *convection* occurs in a fluid volume or between a fluid and a solid, providing a temperature gradient exists. Convection differs from conduction in that the fluid itself can be moving, and thus heat can be transferred not only by "contact" between the molecules, but also by their displacement. If molecular movement is due only to density differences produced, for example, by temperature differences within the fluid, convection is said to be *natural.* In all other conditions, it is described as *forced.* The greater the speed of the fluid in relation to a solid, the greater will be the rate of heat transfer by convection [3].

The general law for this form of heat transfer is similar to that for conduction and can be written

$$\dot{Q}_C = -h_\mathrm{c}\,\Delta T A_\mathrm{c},$$

where $\dot{Q}_C$ is the heat flux exchanged by convection, h_c is the *convective heat transfer coefficient* or *convection coefficient,* ΔT is the temperature difference between the fluid and the solid, and A_c is the surface area transferring the heat.

Convection differs from conduction mainly in the nature of their respective coefficients h_c and h_k. Whereas h_k is dependent principally on the thermal characteristics of the fluid and solid, h_c also depends on such factors as the fluid density p and the relative speed of displacement U.

In thermal physiology, convective heat transfer (C) is defined as the net rate of heat transfer in a moving gas or fluid (i.e., by convection) between different parts of an organism, or between an organism and its external environment; it may develop and be amplified by thermal gradients (natural convection) and by forces such as wind, fans, pumps or body movement (forced convection) usually expressed in terms of unit area of the total body surface [4].

2.4.3.1.2.1 *Natural Convection*

In *natural convection* fluid molecules are moved or displaced only by the influence of density differences created by temperature changes. In its simplest form, consider a fluid with temperature T_∞ in contact to a vertical wall with temperature t_W. In case that the fluid is warmer than the wall, $T_\infty > t_W$, natural convection of the fluid is directed downwards, but upwards in case of the wall being warmer than the fluid, $T_\infty < t_W$ [12].

The convection coefficient h_c is related to the factors involved in this movement:

- The heat conductivity of the fluid, k
- Its viscosity, μ
- Its specific mass (density), ρ
- Its dilation coefficient (thermal expansion coefficient), β
- Its specific heat (at constant pressure), c_p

These factors can be grouped into three *dimensionless numbers* (dimensionless numbers are products or ratios that are expressed without units, in which all constitutive parameters are given in a homogeneous unit system).

The Nusselt number, $\mathrm{Nu} = h_c\, L/k$, is the ratio of convective to conductive heat transfer across (normal to) the boundary.

The Grashof number, $\mathrm{Gr} = g\beta v^2$ approximates the ratio of the buoyancy to viscous force acting on a fluid.

The Prandtl number, $\mathrm{Pr} = \mu c_p/k$, is defined as the ratio of kinematic viscosity, μ/ρ, to thermal diffusivity, $k/\rho c_p$.

In these numbers, L is a geometric characteristic of a body, such as its length (l) or its diameter (d), g is the gravitational acceleration (9.81 $\mathrm{m \cdot s^{-2}}$), and v is the *kinematic viscosity,* which is the ratio of fluid viscosity μ to its density, μ/ρ.

However, whatever the geometric shape of the body and the nature of the fluid, the experimental results show that Nu is a function of Pr and Gr:

$$\mathrm{Nu} = a(\mathrm{Pr} \cdot \mathrm{Gr})^{m},$$

where a and m are constants that can be determined experimentally. Thus, if the above factors and their relationships are known, h_e can be calculated, e.g. in high- or low-pressure atmospheres, or in gas mixtures that differ from air.

2.4.3.1.2.2 *Forced convection*

In *forced convection*, h_c is mainly dependent on displacement of gases or fluids. Therefore, a new dimensionless number is needed to take account of this factor, as the Grashof number relating to free movement of molecules is inapplicable. This new number includes the rate of relative medium (gas or fluid) displacement, U, and is expressed by the ratio

$$\mathrm{Re} = UL/\nu,$$

where Re is the *Reynolds number* and L (geometric characteristics of a body) and ν (kinematic viscosity) have been defined previously.

However, as h_c is dependent on fluid movement, it is also influenced by the patterns of movement and varies according to whether the fluid displacement is laminar or turbulent. In *laminar flow*, Re is low, but in turbulent flow it is high, an approximate transitional zone corresponding to Re values between 2,000 and 10,000. Figure 2.1 illustrates the patterns of laminar and turbulent flow.

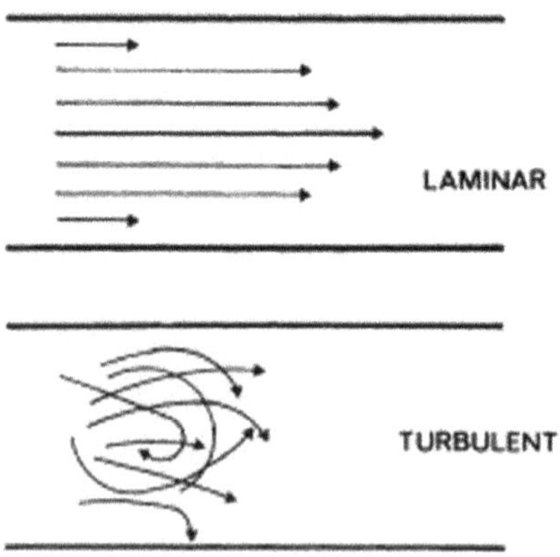

Figure 2.1 Patterns of fluid flow.

In *laminar flow* the fluid can be represented by equal layers sliding easily one on the other. However, if one considers laminar fluid flow along a solid plane, the speed of fluid displacement is not the same close to the surface of the plane as it is at a distance from the surface. The distribution of speed between the value 0 at the interface and the value U at a certain distance x is represented in Fig. 2.2. This flow can be divided into two parts: an external part, where fluid moves homogeneously at speed U, and another part, consisting of a thin layer along the surface within which there is a speed gradient. This layer was called the *boundary layer* by Prandtl at the beginning of the 20th century.

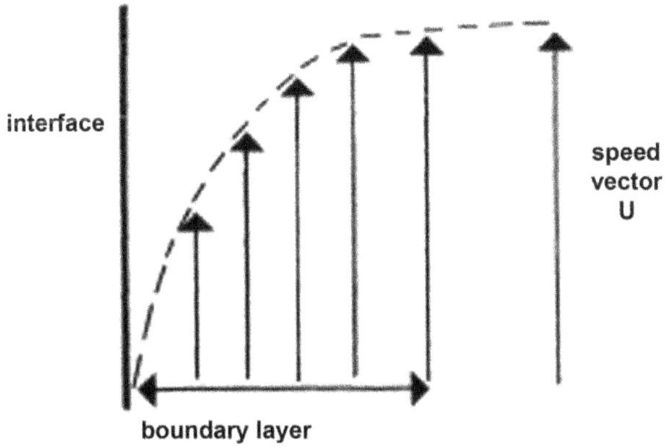

Figure 2.2 The formation of a boundary layer in a laminar flow system.

Turbulent flow is characterized by the existence of local alterations in the flow. Disturbances occur from one point to another, and at a given point from one time to another; therefore, at a fixed point there are speed changes in both intensity and direction. Nevertheless, it can also be assumed that the fluid is composed of at least three successive layers (Fig. 2.3):

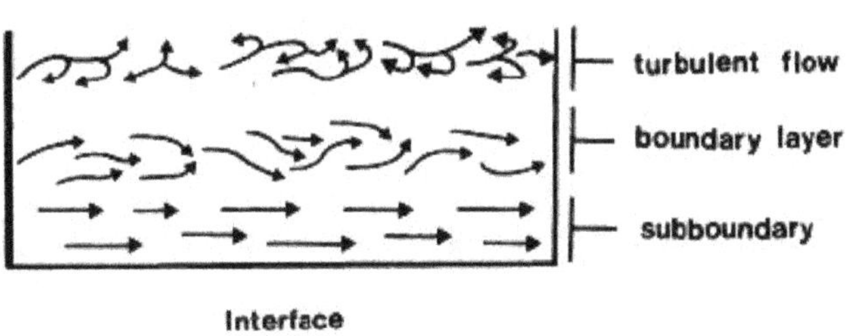

Figure 2.3 Patterns of fluid flow in a turbulent flow system.

1. The flow itself, which is turbulent.
2. A basal laminar flow in close contact with the surface.
3. A barrier layer between the two flows.

In this case, the basal laminar layer is very thin.

By analogy with the concepts of laminar or turbulent flow and of the dynamic boundary layer, it is possible to describe a *thermal boundary layer* in which there is an important temperature gradient between the solid surface and the fluid at a certain distance. At the interface, heat transfer occurs by conduction between the molecules of the solid and those of the fluid. In the fluid itself, heat transfer occurs by true convection.

However, the term *convection* and the related convection coefficient cover the whole phenomenon.

The patterns of flow modify the thickness of the thermal boundary layer and consequently the value of h_c. However, dynamic boundary layers and thermal boundary layers are not of the same thickness. The dynamic interface is thicker than the thermal interface if Prandtl's number is greater than 1, and the inverse is also true. In air, Pr equals 0.72, and the thermal boundary layer is thus thicker than the dynamic boundary layer.

As in natural convection, h_c can be calculated in *forced convection* if Nu is known. Here Nu depends on the characteristics of the fluid expressed by Prandtl's number and also on the patterns of air motion described by the Reynolds number. The general equation is

$$\mathrm{Nu} = a^1(\mathrm{R}^n\mathrm{Pr}^m)$$

where a_1, n, and m are constants. When the fluid is air, Pr is given and the equation may be simplified as follows:

$$\mathrm{Nu} = a_2\mathrm{Re}^p$$

The constants a_2 and p are different according to whether the flow is laminar or turbulent. They can be determined experimentally.

2.4.3.1.3 *Radiation*

All bodies emit electromagnetic radiation (Fig. 2.4) to a greater or lesser degree. In general, *radiant energy*, Q, is the energy travelling in the form of electromagnetic waves (unit: joule, J). This term should be distinguished from the *radiant heat exchange*, R, of the environment with the body [4]. That part of the electromagnetic spectrum of

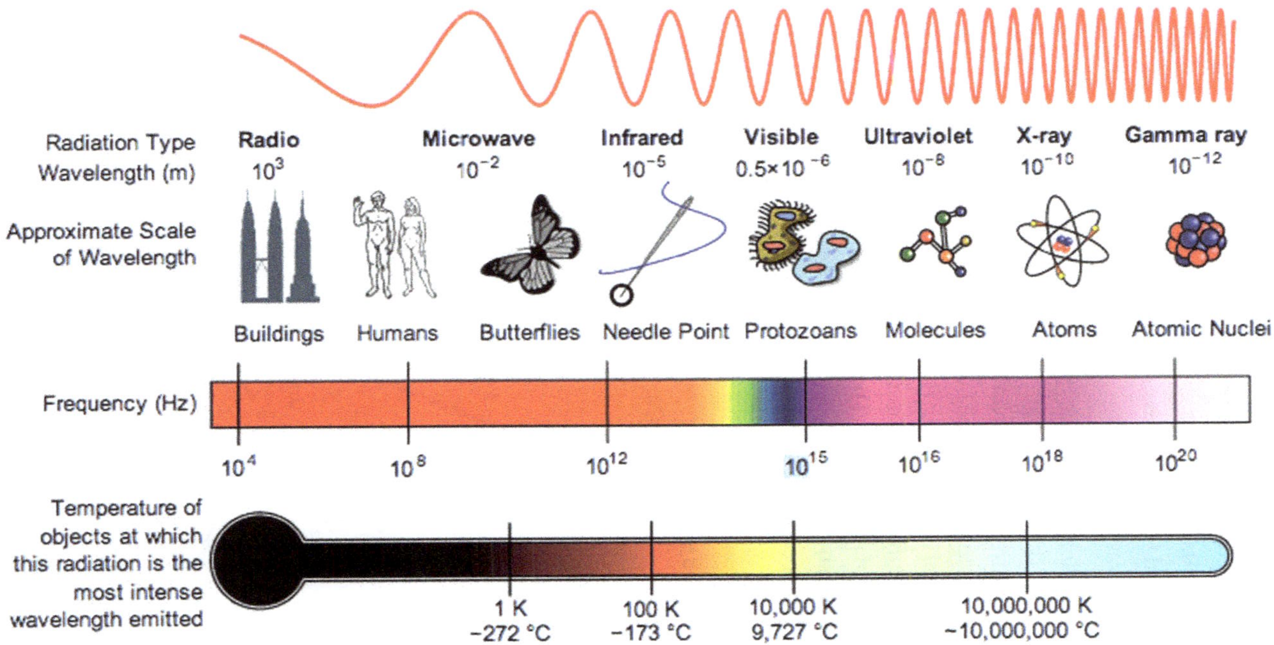

Figure 2.4 Electromagnetic spectrum.

significance in thermal physiology is divided for convenience into the wavebands:

- Ultraviolet 0.25–0.38 μm
- Visible 0.38–0.78 μm
- Infrared 0.78–100 μm
- Microwave 1–100 mm

Certain electromagnetic waves are capable of transporting quantities of thermal energy with wavelengths between 3–30 μm [4] representing thermal radiation. In medicine and biology, mid (3–8 μm) and long (8–15 μm) wavelengths are used for temperature measurement [14]. However, the number and cut-off point of spectral bands in the infrared domain vary in definition of various standards and the field of application. The curve representing the emission of a homogeneously radiating body versus the whole wavelength range is called the *spectroradiometric curve* (Fig. 2.5).

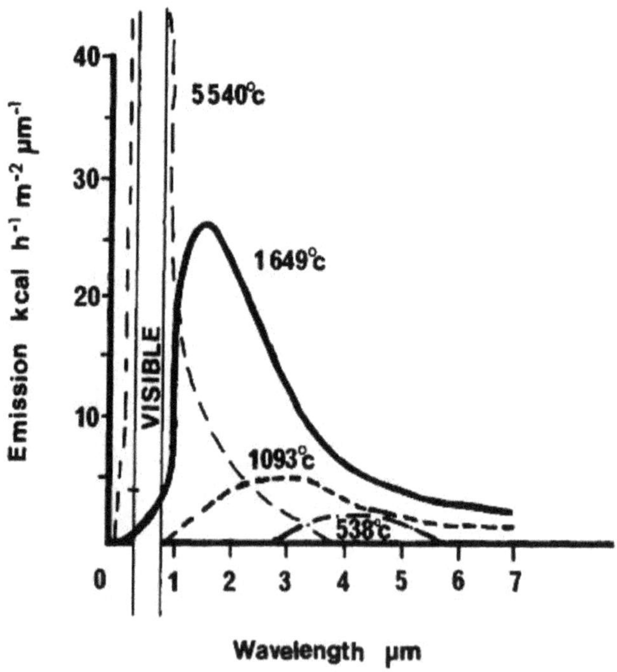

Figure 2.5 The spectroradiometric curve.

2.4.3.1.3.1 Glossary of radiation terms

The "Glossary of Terms for Thermal Physiology" [4] provides a number of terms related to radiation:

Radiance, $L_{e,}$ at a surface element (dA) of a source or receiver is the *radiant intensity* (dI) from direction θ divided by the orthogonal projection of this surface element ($dA \cdot \cos\theta$) on a plane perpendicular to the direction θ; θ is the angle between the normal to the element (dA) of the source or receiver and the direction of the observation (unit: $W \cdot sr^{-1} \cdot m^{-2}$).

Thermal radiance, $L_{e,th}$, is the radiance due to thermal radiation (unit: $W \cdot sr^{-1} \cdot m^{-2}$).

Spectral radiant energy, Q_l, is the radiant energy per unit wavelength interval at wavelength l (unit: $J \cdot nm^{-1}$).

Radiant exitance, M_e, is the radiant flux leaving an element of a surface. This quantity includes radiation emitted, reflected, and transmitted by the surface.

$$M_e = \frac{d_e}{dA} = \int \mathrm{Le} \cdot \cos \Theta \cdot d\Omega \text{ (unit: } W \cdot m^{-2}]$$

The name *radiant emittance* previously given to this quantity is abandoned because it has given rise to confusion. Thus, the term *emittance* is used to designate either the flux leaving a surface whatever the origin (emitted, reflected, or transmitted), or the flux emitted by a surface (originating in the surface), or a quantity without dimensions similar to *emissivity* but applicable only to a specimen.

Radiant self exitance, $M_{e,s}$, describes the *radiant flux* emitted by a surface. The flux considered does not include reflected or transmitted flux (unit: $W\ m^{-2}$).

Radiant thermal exitance, $M_{e,th}$, is defined as the radiant flux emitted as thermal radiation by a surface. In the case of a full radiator, the radiance (L_e) is uniform in all directions. In consequence, when the solid angle is measured in steradians, the *radiant exitance* has the numerical value

$$M_e = 4\pi L_e.$$

Radiant flow, φ, is the amount of *radiant energy* per unit of time (unit:W).

Spectral radiant flow, φ_λ, designates the *radiant flow* per unit wavelength interval at wavelength λ (unit $W \cdot nm^{-1}$).

Radiant flux, dφ/dA, is the *radiant flow* per unit area (unit:W m^{-2}). The radiant flux emitted by a surface is the thermal radiant exitance. The *radiant flux* incident on or passing through an area is the *irradiance.*

Effective radiant flux, H_r, or *effective radiant field* is the net *radiant flux* exchanged with all enclosing surfaces and with any intense directional sources and sinks for exchange of radiant heat by a human or human-shaped object whose surface temperature is hypothetically at *ambient air temperature* (unit $W \cdot m^{-2}$). If the same object, with surface temperature at *ambient air temperature,* is assumed to exchange radiant heat at a given *effective radiant flux* with a hypothetical black surface enclosure of corresponding wall temperature, then this condition defines the effective radiant field transferring the amount of heat per unit time and per unit surface of the object which is quantified by the *effective radiant flux.* In this sense, *effective radiant field* and effective radiant flux are synonymous terms.

Radiant heat exchange, R, is defined by the net rate of heat exchange by radiation between an organism and its environment, usually expressed in terms of unit area of the total body surface, i.e., as a *heat flux.* (unit: W, or $W \cdot m^{-2}$).

Radiant intensity, I, is the radiant flow proceeding from a source per unit solid angle in the direction considered (unit: W sr^{-1}) and *spectral radiant intensity, I_λ,* is defined as the *radiant intensity* per unit wavelength interval (unit: $W \cdot sr^{-1} \cdot nm^{-1}$).

Radiation shape factor, synonym: *radiation view factor, F_{ij},* **is a** dimensionless quantity expressing the fraction of the diffuse energy emitted by a surface (or a source), denoted by the subscript *i*, that is received by another surface, denoted by the subscript *j*, visible by it and in known geometric relation with it.

Emissivity, ε, describes the ratio of the total *radiant energy* emitted by a body to the energy emitted by a *full radiator* at the same temperature.

Directional emissivity, $\varepsilon_{(\theta,\varphi)}$, is the ratio of the *thermal radiance* ($L_{e,th}$) of a body in a given direction to that of a *full radiator* ($L_{e,(\varepsilon = 1)}$) at the same temperature. $\varepsilon_{(\theta,\varphi)} = L_{e,th(\theta,\varphi)}/L_{e,(\varepsilon = 1)}$ in which θ and φ are the angular coordinates defining the given direction.

Hemispherical emissivity, ε_h, is the ratio of the total *radiant energy* emitted by an element of a surface into a hemisphere to the energy by a similar element on the surface of a *full radiator*. The element forms the centre of the equatorial plane of the hemisphere, but it is not necessary to define its radius.

Spectral emissivity, ε_λ, is the ratio of the *radiant flux* emitted by an element of surface per unit wavelength interval to the flux emitted by a *full radiator* at the same temperature and in the same waveband.

Emissivity window, ε_w, is the ratio of the *radiant energy* emitted by an element of surface between wavelength λ_1 and λ_2 to the flux emitted by a *full radiator* at the same temperature and in the same waveband. The quantity is a special type of *spectral emissivity* and is used because some biological materials have a low *emissivity* in parts of the infrared or visible spectrum. These parts are known as windows.

Radiative heat transfer coefficient, h_r, is defined as the net rate of **heat transfer** per unit area by the exchange of thermal radiation between two surfaces, per unit temperature difference between the surfaces (unit: $W \cdot m^{-2} \cdot {}^\circ C^{-1}$).

(Linear) radiative heat transfer coefficient. According to the Stefan–Boltzmann Law, the exchange of radiation between two black surfaces at temperatures T_1 and T_2 (unit: K) is proportional to σT_4, where σ is the Stefan–Boltzmann constant. When the ratio $(T_1 - T_2)/T$ is small, where $T = (T_1 + T_2)/2$, $\sigma(T_1^4 - T_2^4) \approx 4\sigma T^3 \cdot (T_1 + T_2)$ and the term $4\sigma T^3$ can be treated as a linear *heat transfer coefficient* (unit: $W \cdot m^{-2} \cdot {}^\circ C^{-1}$).

2.4.3.1.3.2 *Black body radiator*

Kirchhoff developed the idea of a *black body*, a hypothetical body that sends out radiation of all frequencies and that should therefore also absorb all radiation, and reflect none, so that it appears black [5]. In other words, a *blackbody* or *full radiator* (= emitter of radiant energy) of uniform surface temperature whose *radiant exitance* in all parts of the spectrum is the maximum obtainable. The *emissivity* of a full radiator is unity for all wavelengths, defined as an object that absorbs all radiations falling upon it, at any wavelength.

The emissivity of a blackbody can be described by three equations [3]:

1. *Planck's law.* This general law describes the emitting power, $W(\lambda)_b$ of a blackbody within a spectral interval of 1 μm at wavelength λ

$$W(\lambda)_b = \frac{2\pi hc^2}{\lambda^5 \exp\left(\frac{hc}{\lambda kT}\right) - 1}$$

   where h is Planck's constant (6.6×10^{-34} J . s), k is Boltzmann's constant (1.4×10^{-23} J. K^{-1}), T is the absolute temperature (in K) of the blackbody, and c is the velocity of light (3×10^{10} cm s^{-1}),
2. *Stefan–Boltzmann's law.* This law expresses the total energy emitted by a blackbody, W_b. It was formulated experimentally by Stefan in 1879 and confirmed theoretically by Boltzmann in 1884. It corresponds to the integration of Planck's law from $\lambda = 0$ to $\lambda = \infty$.

$$W_b = sT^4,$$

   where σ is the Stefan–Boltzmann constant (56.7×10^{-9} W . m^{-2} K^{-4}). This law is very important in practical infrared thermography because it states that the total emissive power of a blackbody is proportional to the fourth power of its absolute temperature.
3. *Wien's displacement law.* This law expresses the fact that the wavelength at which the emissive power of a blackbody is maximum varies inversely with the absolute temperature of the body. It corresponds to the derivative of Planck's law with respect to λ:

$$\lambda_{max} = \frac{2898}{T} \text{ (in μm)}$$

This law is the mathematical expression of a common observation: As the temperature of, for instance, an iron bar increases, its colour varies progressively from blue-black to red and yellow, owing to a decrease of the main wavelength of the radiative emission. At room temperature, T is about 300 K and the peak of radiant emittance of common objects lies at 9.7 μm, that is, the far infrared.

2.4.3.1.3.3 *Non-black body radiator (grey body)*

A grey body radiator is a radiator whose spectral emissivity is less than unity, at least in the waveband for thermal radiation (3–30 μm) but is the same at all wavelengths and a *selective radiator* is a radiator with a *spectral emissivity* less than unity which varies with wavelength. Human and animal skins have high *emissivities* in the waveband 3–30 μm but not between 0.7 and 3 μm [4].

Real objects rarely comply with these laws mentioned above over an extended wavelength region, although they may approach the characteristics of a blackbody in certain spectral interval. More typically, when incident radiation falls on a real object, it can be partly absorbed (α), reflected (p), or transmitted (τ) (Fig. 2.6).

Radiation reflectance, ρ, is the ratio of the *radiant flux* reflected by a surface or medium to the incident flux. Measured values of *reflectance* depend upon the angles of incidence and view and the spectral character of the incident flux; these factors should be specified.

Radiation transmittance, τ, is the ratio of the *radiant energy* transmitted through a body to the total radiation incident on it.

Total radiant absorptance, α, is the ratio of *total radiant flux* absorbed by a body to the total incident flux.

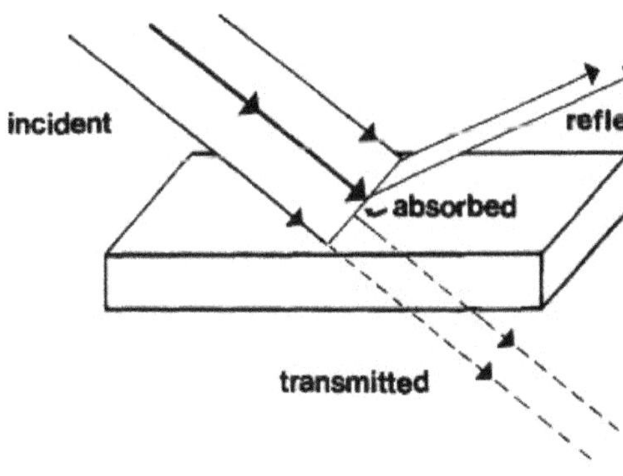

Figure 2.6 The alternative pathways of radiation falling on a surface of another medium.

All these factors are more or less wavelength-dependent. Therefore, the symbol A will be used to imply the spectral dependence of α, ρ, and τ. The sum of these three fractions at any wavelength must be equal to the incident radiation. If the value 1 is attributed to the incident radiation, we have the equation

$$\alpha(\lambda) + \rho(\lambda) + \tau(\lambda) = 1.$$

On the other hand, grey bodies are capable of radiating energy. However, at a given temperature, a grey body always emits less than a blackbody, and its radiant emissivity can be defined by the fraction of its blackbody emissivity. Mathematically speaking, this can be written as the ratio of the object's emissivity, $W(\lambda)_0$ to that of a blackbody $W(\lambda)_b$. If the emissivity of the blackbody is equal to 1, the emissivity of a grey body will vary between 0 and nearly 1.

Kirchhoff's law also applies to black- and non-black (grey) bodies. Therefore, at any wavelength and at any temperature, the spectral emissivity and spectral absorbance of an object are equal:

$$\varepsilon(\lambda) = \alpha(\lambda)$$

We can now define some radiation sources:

The *blackbody* has an emissivity equal to 1 at any wavelength:

$$e(\lambda) = \varepsilon = 1$$

This also means that its absorption is 1 at any wavelength:

$$\alpha(\lambda) = \alpha = 1$$

Therefore, $\rho = \tau = 0$. A blackbody does not transmit any wavelength (i.e., it is *opaque*) and it does not reflect any wavelength.

A *grey body* may be transparent if $\tau(\lambda)$ is different from 0, or opaque if $\tau(\lambda) = 0$. In this case,

$$\alpha(\lambda) + \rho(\lambda) = 1.$$

If the body reflects almost totally the incident radiation, $\rho(\lambda)$ approaches 1. Thus,

$$a(\lambda) = \tau(\lambda) = 0.$$

As $\varepsilon(\lambda) = \alpha(\lambda)$, this body does not emit. A perfect reflecting material such as a mirror or highly polished surface has an emission approaching 0. The emission of a grey body is expressed by the Stefan–Boltzmann law, modified as follows:

$$W(\lambda)_0 = \varepsilon(\lambda) W_b = \varepsilon(\lambda) sT^4$$

2.4.3.1.3.4 *Effect of Surface Characteristics*

Most bodies are non-transparent to infrared radiation. Therefore, only the most superficial layer of the object contributes to the emission of infrared radiation toward the environment. For this reason, the value of T in the Stefan–Boltzmann equation is the temperature of the object's surface. For the same reason, emissivity $\varepsilon(\lambda)$, but also reflectance, is directly related to the fine structure of the surface [15]. For example, iron at 1646 K has an emissivity of 0.25 if its surface is polished, but 0.99 if it is rough. Aluminium alloys are among the materials with a wide variation of emissivity which is most difficult to determine. However, including characteristics of surface roughness improved emissivity prediction of aluminium [15].

2.4.3.1.3.5 *Heat Exchange by Radiation*

Consider a body B_1 which emits radiation toward objects constituting the environment B_2, e.g. the walls of a room. The walls partly absorb this radiation, but they also emit toward the body, which will absorb a fraction of the energy received. The net energy gained by B_1 corresponds to the difference between the energy it has emitted and that which it has absorbed; this is affected by many factors including all the thermal characteristics of the objects and the relative position of the body within its environment.

In practice, this can be described by the following equation:

$$Q_{r=} \sigma \; F_2^1 (T_1^4 - T_2^4),$$

where F_2^1 is a factor embracing the geometric characteristics and various coefficients, such as the emissivity of the bodies. The problem is relatively simple when all temperatures are nearly equal.

According to Wien's law, the main wavelengths will be almost similar. However, if the temperatures are different, the principal wavelengths are different and the coefficient F_2^1 is difficult to determine. Nevertheless, when the temperatures are very different, the problem does become simpler. The best example is that of an object subjected to the sun's radiation. Obviously, the radiation of the object toward the sun is negligible. Thus, the energy transfer is only related to the energy emitted by the sun and to the absorption of the body for the wavelengths of solar radiation.

2.4.3.1.3.5.1 Secondary emission

A body that absorbs heat, for instance from the sun, becomes warmer. It also emits energy through its own infrared radiation. If it becomes warmer, the main emission spectrum will vary. Consequently, solar energy absorbed by an object appears to be reemitted at modified wavelengths. This is the basis for the *greenhouse effect*: Since glass is transparent to solar radiation, this radiation is transmitted to the objects within the greenhouse. Inside the greenhouse, radiation is absorbed by the objects and they then emit radiation at their own temperature, at wavelengths different from those of solar radiation. However, glass is particularly opaque at such wavelengths. Therefore, radiation builds up inside the greenhouse and the temperature increases.

2.4.3.2 Heat transfer with change of state (latent heat transfer)

Latent heat of fusion is defined as the quantity of heat absorbed, or released, in the reversible change of state, melting or freezing, of unit mass of solid or liquid, without change of temperature (unit: $J \cdot g^{-1}$). For example, *the* melting of 1 g of ice to water at 0°C absorbs 333.15 J. The transformation of water into water vapour is only form of heat transfer with change of state that occurs in both, man and animals. Water evaporation (vaporization) requires also energy and the amount of energy needed value is called the *latent heat of vaporization* and symbolized by λ or LHV. It represents the quantity of heat absorbed (or released) by a volatile substance (fluid) per unit mass in the process of its reversible change of state by evaporation (or condensation) under isobaric and isothermal equilibrium conditions (unit: $J \cdot g^{-1}$). For water: $\lambda = 2490.9 - 2.34\ T$, with T = temperature of water in °C, e.g., 1 g of water, when changing from fluid to vapour at T = 34°C, absorbs 2411.3 J. [4].

The net *heat transfer* per unit vapour pressure gradient caused by the evaporation of water from a unit area of wet surface or by the condensation of water vapour on a unit area of body surface is named *evaporative heat transfer coefficient, h_e*. The driving force is the *vapour pressure* gradient from P_{ws} (on the surface) to P_{wa} (of the ambient gas).

Thus,

$h_e = E(P_{ws} - P_{wa})^{-1}$·(unit: W· m^{-2}· Pa^{-1}, or W·m^{-2}·Torr^{-1}).

The evaporative heat transfer coefficient can also be expressed in terms of the *latent heat of vaporization* (λ or LHV) of water, the gas constant for water vapour (R_w), the mean temperature (T) of the medium in K and the *mass transfer coefficient* (h_D)

$$h_e = h_D \cdot \lambda \cdot R_w^{-1} \cdot T^{-1}$$

In most physiological applications, the vapour pressure gradient instead of the concentration gradient can be considered as the driving potential for the evaporative process since the temperature difference between the evaporating surface and the ambient air is small in relation to the average temperature of the surface-water vapour medium.

Mass transfer coefficient (*diffusion*), h_D, is the rate of mass transfer ($\dot{m}$) from a vaporizing liquid (usually water) to a moving gas (usually air) in contact with it, ρ per unit area (A) of the liquid surface and per unit difference between the vapour density (saturated) at the surface (ρws) and the vapour density of the ambient gas (ρwa), expressed in the equation

$$h_D = \cdot \dot{m} \cdot A^{-1} (\rho_{ws} - \rho_{wa})^{-1} \text{ (unit: m s}^{-1}\text{)}.$$

2.4.3.2.1 *The Relationship between mass transfer and convective heat transfer*

The analogy between evaporation and convection means that h_D can be calculated when h_c is known. At this point two new numbers should be considered, namely those of Sherwood (Sh) and of Schmidt (Sc). They are described as follows:

$$\text{Sh} = h_D L/D_w \quad \text{and} \quad \text{Sc} = v/D_w,$$

where L is the geometric expression defining the exchange surface and D_w the coefficient of mass diffusivity of water vapour in air (v has previously been defined as the kinematic viscosity).

In air, Prandtl's number and Schmidt's number are close to unity and consequently close to each other. This means that we also have

$$c_p\mu = k \quad \text{and} \quad v = D_w.$$

Therefore, Nu and Sh are equal. If each is divided by Re, we obtain

$$\text{Nu/Re} = \text{Shi/Re}.$$

It is possible to rearrange this formula in order to obtain

$$h_D = (h_c/\rho_a c_{pa})\ (\mathrm{Pr}/\mathrm{Sc}),$$

which can be simplified into

$$h_D = (h_c/\rho_a c_{pa}),$$

where ρ_a and c_{pa} are the density and the specific heat of humid air respectively. This product varies from 0.25 to 0.32 when one passes from dry air at 0°C to saturated air at 50°C. This relationship is known as the Lewis relation.

By the same reasoning, the coefficient of evaporation may be expressed as

$$h_e = 2.21 h_c$$

if temperature is expressed in °C and pressure in mm Hg. This relationship is very important in practice because it is easy to compute the value of h_e from that of h_c. One therefore arrives at the following important conclusion: If we consider a surface of water A at temperature T, this temperature defines the saturated water vapour pressure P_{ws}^0. On the other hand, the water vapour pressure of the surrounding air is P_{wa} and the driving force is represented by the difference (P_{ws}^0 – P_{wa}). In terms of energy, the amount of energy $\dot{E}$ lost by the water during unit time is expressed by the following equation:

$$\dot{E} = 2.21 h_c (P_{ws}^0 - P_{wa})\, A$$

The value $\dot{E}$ thus corresponds to the maximum quantity of heat that can be lost by vaporization from surface A. It is called the *maximal evaporating power of the ambience.* If the surface is not water itself but only a wet surface, the quantity of water lost into the environment is necessarily less than this maximum value.

References

1. de Boer J, Temperature as a basic physical quantity, *Metrologia* 1965, 1, 158–169.
2. Taylor BN, Thompson A (eds). *The International System of Units (SI) Technology Services*, Natl. Inst. Stand. Technol. Spec. Pub. 330, National Institute of Standards and Technology, Gaithersburg, MD 20899, 2008.

3. Houdas Y, Ring EFJ. *Human Body Temperature. Its Measurement and Regulation*. Plenum Press, New York, London, 1982.
4. IUPS Thermal Commission. Glossary of terms for thermal physiology. *Japanese Journal of Physiology* 2001, 51(2), 245–280.
5. Müller I. *A History of Thermodynamics. The Doctrine of Energy and Entropy*. Springer, Berlin Heidelberg New York, 2006.
6. Lustig RH. Childhood obesity: Behavioral aberration or biochemical drive? Reinterpreting the first law of thermodynamics. *Nature Reviews Endocrinology* 2006, 2(8), 447–458.
7. Taylor NA, Tipton MJ, Kenny GP. Considerations for the measurement of core, skin and mean body temperatures. *Journal of Thermal Biology* 2014, 46, 72–10.
8. Engel J-M, Flesch U, Stüttgen G (eds). *Thermological Methods*. VCH, Weinheim, 1985.
9. O'Sullivan CT. Newton's law of cooling—a critical assessment. *American Journal of Physics* 1990, 58, 956–960.
10. Kleiber M. A new Newton's law of cooling? *Science* 1971, 178, 1283–1285.
11. Narasimhan TN. Fourier's heat conduction equation: History, influence, and connections. *Reviews of Geophysics* 1999, 37(1), 151–172.
12. Lienhard JH IV, Lienhard JH V. *A Heat Transfer Textbook*. 3rd ed, Phlogiston Press, Cambridge, Massachusetts, 2001.
13. Gautherie M. Thermopathology of breast cancer: Measurement and analysis of in vivo temperature and blood flow. *Annals of the New York Academy of Sciences* 1980, 335(1), 383–415.
14. Grenn MW, Vizgaitis J, Pellegrino JG, Percont P. Infrared camera and optics for medical applications. In: Diakides M, Bronzino JD, Peterson DR, eds, *Medical Infrared Imaging: Principles and Practices*, CRC Press, Boca Raton, 2013, 5.1–5.15.
15. Wen C-D, Mudawar I. Modeling the effects of surface roughness on the emissivity of aluminum alloys. *International Journal of Heat and Mass Transfer* 2006, 49, 4279–4289.

Chapter 3

Basic Thermal Physiology: What Processes Lead to the Temperature Distribution on the Skin Surface

List of abbreviations

A_b:	total body area
Ach:	Acetylcholine
A_1, A_2:	adenosine receptors
2-APB:	2-aminoethyl diphenylborinate
ATP:	adenosine-tri-phosphate
AVAs:	arterio-venous anastomoses
$beta_2$, β_2:	beta 2 adrenoceptor
BMR:	basal metabolic rate
°C:	degree Celsius
CPB:	cardiopulmonary bypass
CF:	cystic fibrosis
CFTR:	cystic fibrosis transmembrane conductance regulator
cGMP:	cyclic guanosine monophosphate
CGRP:	calcitonin gene related peptide
CNS:	central nervous system
CESM:	Cold Exposure Survival Model

The Thermal Human Body: A Practical Guide to Thermal Imaging
Kurt Ammer and Francis Ring

ISBN 978-981-4745-82-6 (Hardcover), 978-0-429-01998-2 (eBook)
www.jennystanford.com

E:	evaporative heat transfer
E_{req}:	required evaporative heat transfer
EDHFs:	endothelial derived hyperpolarization factors
EEG:	electroencephalography
EMG:	electromyography
H:	height
$\dot{H}$:	metabolic heat production
h:	heat transfer coefficient
H_1:	histamine-1 receptors
HSP:	Heat shock protein
HSR:	Heat shock response
ICF:	International Classification of Functional Health
IML:	intermediolateral column
IPCC:	Intergovernmental Panel on Climate Change
IR:	infrared
ISO:	International Organization for Standardization
K:	conductive heat transfer
K_{Ca}:	calcium-activated potassium channels
LPB:	lateral parabrachial
LOMR:	lowest observed metabolic rate
M_3:	muscarinic receptor-3
M, MR:	metabolic energy transformation normal metabolic rate
MET:	metabolic equivalent
MMR:	maximum metabolic rate
MOMR:	minimum observed metabolic rate
MVC:	maximal voluntary contraction
NE:	norepinephrine
NK_1:	neurokinin-1 receptors
NO:	nitric oxide
NMDA:	*N*-methyl-D-aspartate
nNOS:	neural NO synthase
eNOS:	endothelial nitric oxide synthase

NPY: neuropeptide Y
PACAP: pituitary adenylate cyclase-activating peptide
PAC_1: pituitary adenylate cyclase-activating peptide 1 receptor
PAG: periaqueductal grey matter
cPAG: caudal PAG
rPAG: rostral PAG
PAS: Periodic acid–Schiff
PCR: Phosphocreatine
PGI_2: prostacyclin
PH: posterior hypothalamus
PMR: peak metabolic rate
POA: pre-optic area
PVN: paraventricular nucleus
R: radiant heat exchange
REHL: respiratory evaporative heat loss
RF: reticular formation
RMR: resting metabolic rate
RPA: raphe/peripyramidal area
RRF: retrorubral field
RVMM: rostral ventromedial medulla
S: storage of body heat
SG: sympathetic ganglia
sGC: soluble guanylate cyclase
SI: système international d'unités
SOED: *Shorter Oxford English Dictionary*
SMR: standard metabolic rate
SP: substance P
T: temperature
T_b: mean body temperature
T_c: core temperature
T_{oe}: oesophageal temperature

T_{sk}:	mean skin temperature
T_r:	rectal temperature
TEF:	thermic effect of food
TNZ:	thermoneutral zone
TRP:	transient receptor potentials
TRPA:	TRP ankyrin
TRPV:	TRP vanilloid
UCP1:	uncoupling protein 1
VIP:	vasoactive intestinal peptide
VMH:	ventromedial hypothalmus
VO_{2max}:	maximum oxygen consumption
$VPAC_1$, $VPAC_2$:	vasoactive intestinal peptide receptor 1 or 2
VTA:	ventral tegmental area
W:	work rate

3.1 Simple Description of Human Physiology

Before we discuss basic temperature regulation, a very simple description of human physiology is provided. "φύσικον (physikon)" was the term used by philosophers in antique Greece to describe the non-living nature. For animals and humans, "φύσις (physis)" was the expression of a living being and physiology is the science of the mechanisms of living creatures or in other words, human physiology is a science discipline that applies physics and chemistry in men to understand the function of the living body. Any living body is constructed of individual cells, which form tissues that are combined to organs and organized in several physiological systems to provide acquirement, distribution and waste management of information, fluid, nutrients and oxygen for energy generation necessary to fuel all above-mentioned processes plus reproduction, regeneration and defence mechanisms, also motion and mobility functions.

Life is an energy-consuming process. The site of energy generation is any individual cell, the principal medium for storage and exchange of energy is adenosine-tri-phosphate (ATP) Living beings need sources of energy which is extracted and transformed from tissues of other living organisms which serve as food. The nutrition system manages within the gastrointestinal tract transport, mechanical and chemical processing of food and fluids, the resorption of chemical compounds such as carbohydrates, lipids, proteins, water and minerals and provides the excretion of the non-usable rest of food.

Water and minerals are recycled in the kidneys and water is excreted via the lower urological tract. Nutrients are transported through the walls of the upper intestinal tract into the vascular system which distributes fluid, nutrients and hormones to all tissues of the organism. Hormones are proteins which control the metabolism of target cells and initiate special cell functions. The hormones of the thyroid gland modify the activity of basic cell metabolism.

Another important chemical element necessary for humans and animals is oxygen, which is inhaled with air, transported through the gas-exchanging surface of the lung to the capillary net adjacent to the pulmonary alveoli, bound to haemoglobin of red blood cells, which act as oxygen carrier and is delivering oxygen to the cells of all tissues. The principle to release internal energy of chemical compounds through oxidation was already mentioned in Chapter 2. Measuring the uptake of oxygen and/or the elimination of carbon dioxide, and the conversion of these values to an equivalent quantity of heat is the most common method of indirect calorimetry. The calorific equivalent of a litre of oxygen is approximately 20 kJ of heat. Oxygen consumption has a key role in energy metabolism, which is defined as the sum of the chemical changes in living matter in which energy is transformed. The maximum rate at which an organism can take up oxygen is termed maximum oxygen consumption, VO_{2max}, expressed in SI units as $ml \cdot s^{-1}$, but conventionally ml per min and l per min are the used units. Determination of this parameter requires very high motivation of the subject and can probably be done only on humans. Criteria used to show that a human subject has reached the VO_{2max}, although not yet agreed upon, include an indication of no further increase in oxygen uptake during further increase in workload. More details of metabolic heat generation will be described later.

Another important function of a living organism is processing and distribution of information. Sensing and transport of information either arriving at the body surface or generated inside of body structures, is the task of the peripheral sensory and autonomic nerve system. The processing of this information takes place in the spinal cord, and the brain. While electrical current achieves the speed of information in nerves, this current is generated due to the exchange of ions across the cell membrane of nerve fibres. These electrical phenomena are coupled to the release of a chemical messenger (neurotransmitter) from the nerve cell. Once released, this chemical messenger can diffuse across the gap to the target cell, where it can bind and interact with a specific protein (receptor) embedded in the cell membrane. This process of binding leads to a series or cascade of secondary effects which result either in a flow of ions across the cell membrane, or in the switching on (or off) of enzymes inside the target cell. A biological response then results, such as the contraction of a muscle cell or the activation of fatty acid metabolism in a fat cell [1].The structure of ion channels is in focus of intensive research. These pores in the cell membrane are generated by proteins, which are expressed under the influence of chemical, electrical or temperature stimuli.

Cells receive also chemical messages from circulating hormones. Once again, receptors are responsible for binding these messengers and triggering a series of secondary effects. The action of chemical messages is always the same, irrespective of the way the messenger travelled to the target cell. There are a large variety of messengers, that interact with receptors and they vary significantly in structure and complexity. Some neurotransmitters are simple molecules, such as monoamines (e.g. acetylcholine, noradrenaline, dopamine, and serotonin) or amino acids.

Nerve signals to the adrenal gland may result in the release of stress hormones such as cortisol and adrenaline. (Nor)-Adrenaline has multiple functions. Besides its action as hormone, which evokes protein and enzyme production in target cells when bound to their surface, noradrenaline is the main neurotransmitter in peripheral sympathetic nerve fibres.

It is important to understand the mechanisms of control in physiological processes. The "Glossary of Terms for Thermal Physiology" [2] provides useful definitions of the terms control and regulation. All examples related to thermoregulation have been removed from these definitions below and the application of the general principle of regulated physiology to thermoregulation will be discussed later in more details.

Control

1. The process by which a physiological variable becomes a function of information generated locally or transmitted from remote sources by neural or hormonal signals (feed forward control)
2. The process by which a physiological system stabilizes ("regulates") a variable, generally by means of an information loop with negative feedback. The control loop comprises the controlled (passive) system and the controlling (active) system, which may be divided into sensor, processor and effector components.

With the adoption of definition #2 from the terminology of technical control, further terms like set point, controlled variable, load-error, negative-feedback, servo-control, etc., have been introduced to describe the various components of the physiological processes of regulation. Control in the sense of definition #2 and regulation are analogous terms in physiology. However, the use of the term regulation should be generally given preference to avoid confusion between the two definitions of control.

Control, proportional

A control concept with negative feedback, in which the controller activates effector mechanisms to an extent dependent on the deviation of the controlled variable from its set point. If the disturbance persists, such a control concept implies a load error, i.e. a permanent deviation of the controlled variable. However, such a deviation is much smaller than that which is present without feedback control.

Regulation

The processes by which a biological system stabilizes variables is generally using of information loops (negative feedback control). The regulated (controlled) physiological variables can usually be described as a function of physical variables. In the case of changes of a regulated variable due to external or internal interference, the processes of regulation will alter certain bodily functions, the effectors in the regulatory process, which act to minimize the deviations of the regulated variable from its set point.

3.2 Simple Description of Temperature Regulation

Although body temperature measurements started to become a regular medical procedure in the middle of the 19th century, comprehensive models of thermoregulation were developed in the 50s and 60s of the 20th century. Houdas and Ring described the state of art of the time in an easy understandable way when they published their book *Human Body Temperature* in 1982 [3]. In the 1990s, evidence was raised that sympathetic nerve fibres control both vascular skin responses to thermal stimuli, vasodilation in case of heat load and vasoconstriction in case of heat loss. Recently the model of a single central temperature control was rejected and exchanged with several control loops integrating deep and superficial body temperatures [4]. However, the traditional temperature regulation model described by Houdas and Ring is still useful to understand the principle of human thermoregulation.

3.2.1 Anatomical and Physiological Systems

It became obvious from early temperature measurements in men and animals that the distribution of temperature is not homogeneous in living bodies. Temperature distribution differs between species as different adaptation to the temperature environment of living beings have emerged. One group are homeotherm animals such as mammals, which can maintain their deep body temperature within a narrow range. This became possible through physiological processes, which actively counteract the equilibration between the heat energy stored in the living body and the heat content of the environment.

Temperature regulation is an important physiological subsystem, functioning in close relationship with the circulations system using the vascular network to transport heat by conduction from deep body tissue to the surface. There is also a strong dependence from the system that controls fluid haemostasis, as sweating is an important mechanism for dissipating heat. The autonomic nerve system controls the smooth muscles of the vascular walls, changing in that way the inner width of blood vessels and thus controlling blood flow and blood distribution. The purpose of thermoregulation is to keep the core temperature in a narrow range irrespective of external or internal conditions add heat or remove thermal energy from the body.

For maintaining the heat content of an object exposed to external heat load or heat loss, two possible mechanisms are available. One is to modify the heat transfer to the environment by changing the thermal resistance and/or heat dissipation, the other is increasing or decreasing heat production. For both solutions, a regulation loop is necessary, constructed of sensors, a processor unit and effectors. Figure 3.1 shows the components of the autonomic temperature regulation loop. Information from various sensors is fed to the central processing, the output of which goes to the thermo-effectors. These thermo-effectors promote either production or dissipation of heat and keep in that way the deep body (core) temperature relatively constant resulting in temperature gradient directed from deep tissues to the body surface that separates the controlled temperature core from the temperature shell. The later varies in both heat intensity and extension of the temperature field.

It must be noted that regulation of deep body temperature by a unified system with a single controller is now an obsolete model. The current model proposes that deep body is regulated by independent thermo-effector loops, each having its own afferent and efferent branches. The activity of each thermo-effector is triggered by a unique combination of shell and core temperatures.

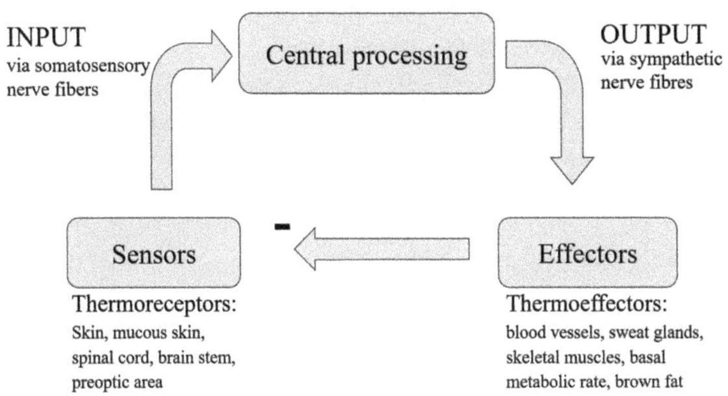

Figure 3.1 Thermoregulation loop.

3.2.1.1 Information line: Thermosensors

Thermoreceptors are located either in the skin or in the body core, especially in the central nervous system (CNS]. They are termed *peripheral thermoreceptors* and *central thermoreceptors,* respectively.

The "Glossary of Terms for Thermal Physiology" defines a thermoreceptor (synonym: temperature receptor) as thermosensitive neural element for which both its afferent function and its response characteristics are electro-physiologically identified. Thermoreceptors have been unequivocally identified, so far, only in the skin and mucous surfaces as cold receptors, warm receptors, and infrared receptors in some snake species, vampire bats and insects including forest fire-seeking beetles, certain butterflies and blood-sucking bugs [5].

The definition of a thermosensor is a neural element or circuitry of neural elements for which it is established by psychophysical criteria or analysis of thermos-effector responses or changes of core temperature that they transduce temperature in such a way that thermal sensation is elicited and/or temperature regulation is adequately stimulated. Thermosensors may not (yet) be accessible to electrophysiological identification, but their function is equivalent to that of thermoreceptors which are electro-physiologically identified thermosensors.

The glossary also differentiates between thermo-reactive, thermo-responsive and thermosensitive.

Thermo-reactive is the descriptive of neural elements whose activity changes with the temperature of a remote region of the body, due to synaptic input from this region. A central nervous neuron may be both thermo-reactive and thermo-responsive (thermosensitive).

Thermo-responsive (synonyms: temperature responsive, thermos-sensitive) is a neural element which changes its activity in response to changes of its own temperature. In comparison to the more general term temperature dependence, neuronal thermo-responsiveness implies at least the possibility of generation or modulation of information by temperature, either non-specific or as a specific temperature signal. The synonymous term thermosensitive (synonym: temperature sensitive) implies the specificity of the thermo-responsive neural element as a temperature signal generator.

The electrophysiological characteristics of thermoreceptors are established since the 1980s, but the molecular basis of cold and warm receptors was elucidated in the beginning of the 21st century by the detection of transient receptor potentials (TRP) [7].

Six thermo-transient receptor potentials (TRPs) have been characterized to date: TRP vanilloid (TRPV) 1 and 2 are activated by painful levels of heat with thresholds ≥42°C and ≥52°C, respectively. TRPV 1 can be expressed by capsaicin, the active ingredient in chilli peppers, also by EtOH allicin, an alcoholic extract from garlic. Camphor, low (acidic) pH-value, lipoxygenase, resiniferatoxin and an andamide act as agonists to temperature on TRPV1 expression. TRPV1 can be found in peripheral thermoreceptors located in the skin, tongue and the peripheral nerve system, but are also available in the bladder, the spinal cord and the brain.

TRPV 2 is widely expressed in both the peripheral and central nervous system.

TRPV3 and -4 respond to non-painful warmth. TRPV3 with a threshold ≥33°C is activated by camphor and 2-aminoethyl diphenylborinate (2-APB) and was in keratinocytes and human sensory neurons.

TRPV4, originally identified as an osmo-sensory ion channel, is also activated by temperatures between 27°C and 42°C. Phorbol esters and low osmolarity lead to expression of TRPV4, which was identified in the kidney, liver, trachea, heart, skin, fat, peripheral and central nervous system including the inner ear, hypothalamus and brain.

TRP melastatin (TRPM)8 is activated by non-painful cool temperatures, while TRP Ankyrin (TRPA)1 is activated by painful cold [5]. In general, innocuous cool is defined as temperatures between 30°C and 15°C, whereas noxious cold is generally perceived as temperatures below 15°C.

TRPM8 is activated by temperatures <25°C and chemically by two natural compounds, menthol which is extracted from mint, and eucalyptol, main compound in eucalyptus oil, and the synthetic compound icilin. TRPM8 was identified in the human prostate and the peripheral nervous system of mammals.

TRPA1 was localized in the peripheral nervous system and in the hair cells of the inner ear. A temperature less than 17°C activates the transient receptor potential thermally, cinnamaldehyde, main compound of cinnamon essential oil, mustard oil, allicin and icilin lead also to the activation of TRPA1.

The mechanisms that underlie thermal activation of TRP channels are not completely understood. However, differences in heat capacity of open and closed forms of TRP channels may be functionally important.

Since all neurons show some sensitivity to temperature, by definition, a *thermosensitive neuron* exhibits some specialized properties. The best evidence for this is founded on the study of the Q_{10} factor. The "Glossary of Terms for Thermal Physiology" provides the following definition of Q_{10}: The ratio of the rate of a physiological process at a defined temperature to the rate at a temperature 10°C lower, when the logarithm of the rate is an approximately linear function of temperature.

The most ion channels in neurons show small linear increases in current flow with Q_{10} values lower than 2. A Q_{10} greater than 2 denotes a special sensitivity to temperature and is related to the property of thermoreception, for example, TRPV1, a very efficient receptor for noxious heat, has a Q_{10} value of 20.6.

As thermoreceptors are proteins in structure, other molecules such as intracellular and extracellular proteins, change of pH-value, can activate them change in membrane voltage and of course by temperature. A device named a thermode, which can deliver controlled heat intensities at the contact side of the device, is used for temperature stimulation of both peripheral and central thermoreceptors.

3.2.1.1.1 *Peripheral thermoreceptors*

Spatially separated sites in the human skin where different intensities of heat can be sensed are known since 1883 [6]. We now term these points, which are the free endings of nerve fibres, unmyelinated C-fibres or thinly myelinated Aδ-fibres as peripheral thermoreceptors, and they represent the peripheral sensors in the thermoregulation loop.

Peripheral thermoreceptors provide input for two physiological functions:

(1) an afferent temperature signal for thermoregulation which maintains deep body temperature within a narrow range through activation of appropriate thermo-effectors

(2) the capability to detect potentially noxious thermal stimuli that pose an immediate threat to the integrity of the integument (i.e. noxious cold and heat stimuli)

For better understanding of the thermal response profile of skin thermosensors, some terms of thermal and neuro-physiology must be explained.

Thermoneutral zone (TNZ) is defined as the range of ambient temperature at which temperature regulation is achieved only by control of sensible heat loss, i.e. without regulatory changes in metabolic heat production (*H*) or evaporative heat loss. The thermoneutral zone (TNZ) will therefore be different when insulation, posture or basal metabolic rate (BMR) varies. The thermoneutral zone (TNZ) should be distinguished from thermo-effector threshold zone (inter-threshold zone), which is defined in Section 3.2.1.3.

In neurophysiology, adaptation is defined as a process of reduced neuronal response to a continuous exposure to stimuli. This must not be confused with adaptation in thermal physiology that is described as changes that reduce the physiological strain produced by stressful components of the total environment. This change may occur within the lifetime of an organism (phenotypic) or be the result of genetic selection in a species or subspecies (genotypic). In thermal physiology, the use of the term adaptation does not require specification of the climatic components of the total environment to which the organism adapts, but the most obvious component is often denoted (e.g. adaptation to heat). There are no distinct terms that relate genotypic adaptations to the climate or components of climate. In comparison to adaptation as defined in neurophysiology, the adaptive processes in thermal physiology usually occur with larger time constants.

Cutaneous thermoreceptors typically exhibit one of four different thermal response profiles.

The first type of thermal response is related to the detection of noxious cold. These fibres are generally silent at the thermoneutral temperature, exhibit thermal thresholds at around 20–10°C, show a linear increase in firing intensity upon cooling down to 0°C and exhibit limited adaptation.

The second type of response is observed in fibres activated by cool temperatures of 37–20°C. Most of these fibres exhibit ongoing activity at thermoneutral skin temperature, which disappears upon moderate warming. Discharge frequency increases upon moderate cooling, reaching a maximum firing rate between 30°C and 20°C. At temperatures of around 20°C, fibres generally show rapid adaptation, which leads to a reduction in discharge frequency at temperatures below ~17°C.

A similar bell-shaped temperature–frequency relationship and rapid adaptation is observed in warm-sensitive fibres. These exhibit continuous activity at thermoneutral skin temperature, which disappears upon moderate cooling. Their discharge rate increases upon moderate heating, reaching a maximum at 40–43°C, but these fibres rapidly fall silent upon further heating.

Finally, fibres involved in sensing noxious heat are typically activated at temperatures above 43°C, with peak discharge occurring at noxious temperatures (45–53°C) and show little or no adaptation.

In the cat, some peripheral thermoreceptors may react to slight changes of skin temperatures such as an increase of 0.3°C or a decrease or a decrease of 0.2°C [8].

3.2.1.1.2 *Central thermoreceptors*

The term central is here used as the opposite of peripheral and implies only that thermal receptors have been found in other parts of the body. They are most numerous, however, in the CNS, particularly in the hypothalamus.

3.2.1.1.2.1 *Hypothalamic thermosensitive neurons*

The hypothalamus contains some neurons that respond by modifying their impulse discharges when their temperature is changed, and these receptors are termed thermo-responsive or thermo-sensitive. If these neurons respond exclusively to stable and changing temperature, they have been termed true or primary thermoreceptors [3]. About 40% of neurons in the hypothalamus are thermo-sensitive.

Such neurons may also receive and transmit a nervous signal related to remote temperature variation and termed thermo-reactive. On the other hand, there are in the hypothalamus some neurons that produce a thermal effector reaction when they receive electrical, but not thermal, stimulation. These neurons bear exclusively thermo-reactive functions and are considered part of the thermoregulatory centres.

3.2.1.1.2.2 *Mesencephalic thermal receptors*

Research in the 1970s reported that some neurons of the mesencephalic reticular formation exhibit a high thermal sensitivity. According to the criteria given above, 8% of these neurons can be considered as cold primary detectors and 79% as warm primary detectors, whereas the remaining 13% are interneurons. Histologically they have a wide distribution within the mesencephalic reticular formation dorsal to the red nucleus. Thermo-responsive neurons have been identified in the caudal para-median midbrain of cats, and a much smaller percentage in rostral regions of the midbrain. In the rat, about 30% of midbrain neurons were thermo-reactive to temperature changes in the pre-optic region.

Recently, thermo-sensitive and thermo-reactive serotonergic neurons were detected in the rat dorsal raphe nucleus and raphe pallidus nucleus using immune-histochemical detection of the protein product of the immediate-early gene c-fos.

3.2.1.1.2.3 *Thermal receptors in the spine*

Many experiments show that cooling or warming the spinal cord in un-anaesthetized animals induces thermoregulatory responses. It has been demonstrated that these responses they do not depend on the stimulation of efferent fibres but are due to the stimulation of primary thermoreceptors. These spinal receptors exhibit static and dynamic responses, as do the other true receptors. Recently [10], the information transfer from spinal thermoreceptors to CNS was described as follows: Spinal cord temperature contributes physiologically to temperature regulation. Parallel anterolateral ascending pathways transmit signals from spinal cooling and spinal warming: they overlap with the respective axon pathways of the dorsal horn neurons that are driven by peripheral cold- and warm-sensitive afferents. By a presynaptic action on their central terminals, local spinal cooling enhances neurotransmission from incoming "cold" afferent action potentials to second order neurons in the dorsal horn; this effect disappears when the spinal cord is warm. Spinal warm sensitivity is due to warm-sensitive miniature vesicular transmitter release from afferent terminals in the dorsal horn: this effect is powerful enough to excite second order neurons in the "warm" pathway independently of any incoming sensory traffic.

3.2.1.1.2.4 *Other central thermoreceptors*

Thermosensitive TRPs have been identified in the inner organs of some animals including the bladder (TRPV1), in the kidney, liver, trachea, heart, fat tissue (TRPV4, which acts also as osmo-sensory ion channel) and in the human prostate (TRPM8).

Thermosensors in the intestine of humans, sheep and rat may trigger thermoregulation responses. Some of these thermosensitive neurons are mechanoreceptors and respond to temperature stimuli by changing the intestinal tone. TRPV1 is widely expressed in peripheral terminals throughout the gut; however, the temperature characteristics of this vanilloid receptor question its role of an intestinal thermo-sensor.

It has also been suggested that the walls of large veins and muscle fibres could contain thermosensitive cells, but evidence for this location is conflicting.

3.2.1.2 Information line: Pathways

3.2.1.2.1 *Afferent pathways*

Temperature information is finally transmitted to the somatosensory cortex, where the perception of cold and warm is generated. Temperature sensing is also connected with nociception and elicits autonomic thermal avoidance responses, and most important, initiates thermoregulatory responses (Fig. 3.2).

Cutaneous primary sensory neurons involved in thermal sensation (green) include both non-myelinated C fibres and thinly myelinated Aδ fibres. The cell bodies of these neurons are in the dorsal root ganglia (DRGs) and have axons with two branches. One branch extends towards the periphery, with free endings in the skin, where thermal information is transduced by TRPs and coded in the form of action potentials. These action potentials propagate to the end of the other axonal branch, which forms synapses in layers I and II (for C fibres) or layers I and V (for Aδ fibres) of the dorsal horn. Activity of both Aδ and C primary sensory fibres controls three distinct neuronal pathways and ensuing responses: First, motor neurons can be activated via spinal interneurons, leading to a rapid withdrawal reflex in response to noxious temperatures (red neurons); second, thermo-sensory information is transmitted via second-order sensory neurons of the ascending spinothalamic tract to the thalamus and further relayed to the somatosensory cortex, where our perception of temperature is formed (blue neurons on the coronal section of the human brain); and third, thermosensory information is transmitted via lateral parabrachial (LPB) neurons, which may also receive input from the spinothalamic neurons, to the pre-optic area (POA) of the hypothalamus (blue neurons on the central sagittal section of the human brain), where thermoregulatory processes are initiated.

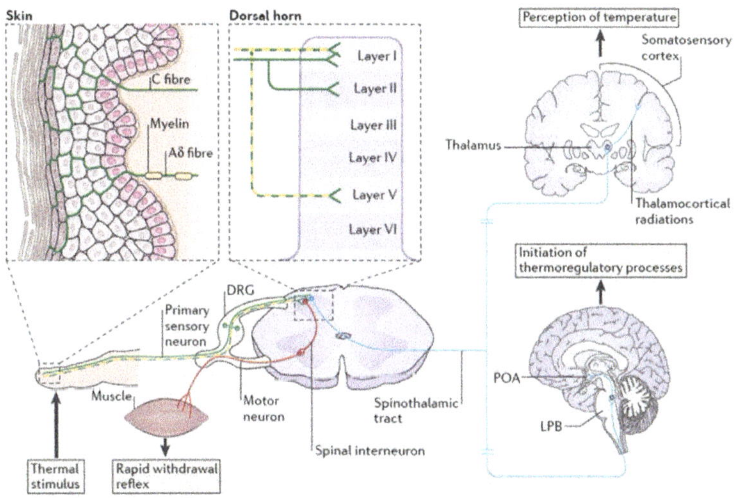

Figure 3.2 Pathways of temperature information transmission.

3.2.1.2.2 *Efferent pathways*

The information to autonomic effectors such as control of skin vessels, non- shivering and shivering thermogenes is travels over different routes. This also applies to behavioural effectors, although little is known about the neuroanatomy of behavioural thermoregulation [9].

The information for controlling the skin vasculature is originated from the response of warm-sensitive cells located in the pre-optic area (POA) of the hypothalamus. These cells have branches that excite cells of mesencephalic peri-aqueductal grey matter (PAG), while other branches inhibit neurons located in the ventral tegmental area (VTA) of the midbrain. Axons of the midbrain neurons descend the spinal cord with the reticulospinal and rubrospinal tracts to cells located primarily in the raphe/peripyramidal area (RPA) of the medulla. Projections from VTA neurons towards the RPA are excitatory, but inhibitory from PAG neurons. The nerve signal from RPA is transmitted to preganglionic neurons with bodies located in the inter-mediolateral column of the spinal cord. Postganglionic neurons of the regional sympathetic ganglia finally control the smooth musculature of skin vessels.

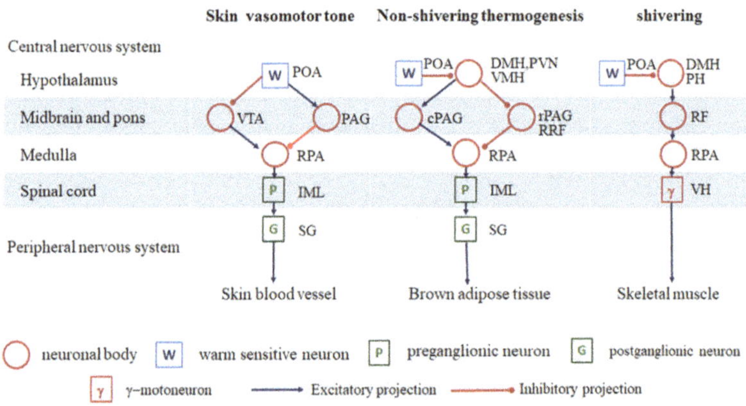

Figure 3.3 Efferent neuronal pathways for control of skin vasomotor tone, non-shivering thermogenesis in brown adipose tissue, and shivering in the rat (after [9]). *Abbreviations*: DMH, dorsomedial hypothalamus; IML, intermediolateral column; PAG, peri-aqueductal grey matter; cPAG, caudal PAG; rPAG, rostral PAG; PH, posterior hypothalamus; PVN, paraventricular nucleus; RF, reticular formation; RPA, raphe/peripyramidal area; RRF, retrorubral field; SG, sympathetic ganglia; VH, ventral horn; VMH, ventromedial hypothalamus; VTA, ventral tegmental area.

Remarkably little is known about of the central pathway that activates sweating. In the cat, sympathetic premotor neurons for sweating may reside in the juxtafacial region of the rostral ventromedial medulla (RVMM) [11].

In humans, the neural pathway from the brain to sweat gland is thought to be as follows; efferent signals from the pre-optic hypothalamus travel via the tegmentum of the pons and the medullary raphe regions to the intermediolateral cell column of the spinal cord. In the spinal cord, neurons emerge from the ventral horn, pass through the white ramus communicans and then synapse in the sympathetic ganglia. Postganglionic non-myelinated C-fibres pass through the grey ramus communicans, combine with peripheral nerves and travel to sweat glands [12].

For non-shivering thermogenesis, the neural pathway from RPA cells to the target tissue is identical to the route of controlling skin vasomotor tone, but the pathway from the hypothalamus to the RPA is different. Cold sensitive neurons are in the dorsomedial (DMH) and ventromedial (VMH) hypothalamus and in paraventricular nucleus (PVN). They receive inhibitory input from hypothalamic warm receptors in the POA and from the mesencephalic retrorubral field (RRF) and the rostral periaqueductal grey matter (rPAG). DMH, VMH and PVN neurons excite cells of the caudal periaqueductal grey matter (cPAG) that finally activate cells in the raphe/peripyramidal area (RPA) of the medulla [9].

Within the shivering pathway, γ and α motoneurons of the ventral horn receive direct and indirect inputs from the midbrain and brain stem, including the raphe/peri-pyramidal area of the medulla.

3.2.1.3 Effectors

The "Glossary of Terms for Thermal Physiology" defines thermo-effectors in relation to body heat balance. A thermo-effector is an organ system and its action, respectively, that affects heat balance in a *controlled* manner as part of the processes of temperature regulation. A multitude of thermo-effectors is involved in both autonomic and behavioural temperature regulation.

Thermo-effector gain is defined as the derivative of the thermo-effector output with respect to body temperature deviation from the set point. The gain may change due to fever, adaptation and other processes. In combination with thermo-effector threshold temperature, thermo-effector gain is an important characteristic of thermo-effectors.

Thermo-effector threshold is the level of activity of a potential thermo-effector that is transgressed when it becomes actively involved in temperature regulation.

For some of the thermo-effectors, thresholds can be precisely determined, e.g. as the level of basal metabolic rate (BMR)or resting metabolic rate (RMR)of a tachy-metabolic animal above which metabolic heat production ($\dot{H}$) will increase in response to a sufficiently strong cold exposure. The thresholds of other thermo-effectors are arbitrarily defined, or agreed upon by convention, because basal or resting levels are difficult to define, e.g. in case of circulatory convection of heat to the skin and the underlying cutaneous vasomotortone.

Thermo-effector threshold temperature describes the level of a specified body temperature (e.g. core temperature or mean body temperature) the transgression of which in one direction, either upward or downward, will activate a certain thermo-effector. As a rule, the threshold core temperature determined for a given effector will be a function of skin temperature, and vice versa; similar interdependencies exist with and among the threshold temperatures determined for specified thermo-sensor regions in the body core (e.g. hypothalamus, spinal cord). Evaluation of mean body temperature threshold values tries to account for the entire thermal input from the body. Specified threshold temperatures for certain thermo-effectors may change during the processes of fever, acclimatization and adaptation. Threshold temperature and thermo-effector gain are the dominant parameters describing the thermo-effector output as a function of body temperature, thus, they constitute essential characteristics of the thermal controller.

Thermo-effector threshold zone (inter-threshold zone) is the temperature range between two threshold (body) temperatures, for activation of any thermo-effector responses, particularly of metabolic heat production ($\dot{H}$) and of evaporative heat loss when no thermal load is present. This special steady state may be called set point. Thermo-effector threshold zone should be distinguished from thermoneutral zone (TNZ).

The efferent branches of thermoregulation loops control various autonomic effectors that are activated in a stepwise manner.

The first step of counteracting a disturbance of deep body temperature is regulation of the blood flow in skin vessels via control of the vascular musculature. Cold skin temperatures lead to vasoconstriction and warm skin results in vasodilation.

Severe heat loads provoke the additional heat defence mechanism of sweat evaporation on the skin surface, while exposure to cold environment activates non-shivering and shivering heat production.

More details of these mechanisms will be provided in Sections 3.5.1 and 3.5.2.

3.2.1.4 Temperature regulation

The direction of heat transfer between two objects is dependent on the difference of heat energy intensity and the distance in material between the objects. Heat may be generated by direct transformation of any other form of energy or as by-product of incomplete transformation of one form of energy into another. For either heat sources, a homogeneous distribution of uniform intensity is unlikely. That means each object which produces heat autonomously or receives heat from an external source, has a core region of temperature surrounded by a temperature shell dependent on the environmental temperature outside of the heated object. Material characteristics of an object predict the distribution of a defined amount of heat. Heat will expand along the material which has a given thermal conductivity which is reciprocal to thermal resistance, defining in that way the volume of both core and shell temperature.

Thermoregulation aims to maintain the core region of the temperature field within a narrow range irrespective of internal or external head gain or heat loss. That means that prior to final equilibration, the natural inhomogeneity of heat intensity within a body will be preserved due to thermoregulatory processes.

The "Glossary of Terms for Thermal Physiology" provides some general definitions related to thermoregulation, which have a wide scope and are not restricted to the thermoregulation system in humans. As many results in human thermal physiology are derived from animal experiments, it may be helpful to get a slight idea of the variations in thermoregulation in different species.

The general term is temperature regulation. Two mechanisms can be differentiated, i.e. autonomic and behavioural thermoregulation. Older now obsolete terms are chemical and physical thermoregulation and physiological thermoregulation when used as synonym for autonomic. Consequently, temperature regulators and temperature conformers are differentiated. The definitions of terms are shown in Table 3.1.

Table 3.1 Terms related to temperature regulation [2]

Temperature regulation

The maintenance of the temperature or temperatures of a body within a restricted range under conditions involving variable internal and/ or external heat loads. Biologically, the existence of body temperature regulation is to some extent by autonomic or behavioural means. Antonym: temperature conformity.

Temperature regulation, autonomic

The regulation of body temperature by autonomic (i.e. involuntary) thermo-effector responses to heat and cold, which modify the rates of heat production and heat loss (i.e. by sweating, thermal tachypnea, shivering, non-shivering thermogenesis, and adjustments of circulatory convection of heat to the surfaces of the body). (Greek. αυτοσ *autos*-self; νομοσ-*nomos*-law, i.e. self-governing, from *Shorter Oxford English Dictionary* (*SOED*)).

In this definition, the term *autonomic* is used in its more general sense and does not imply that all responses are controlled by the *Autonomic Nervous System* (sympathetic and parasympathetic efferents). Autonomic temperature regulation is frequently described as physiological temperature regulation, a term which should be used for both autonomic and behavioural thermoregulatory processes.

Temperature regulation, behavioural

Any coordinated movement of an organism ultimately tending to establish a thermal environment that represents a preferred condition for heat exchange (heat gain, heat loss, or heat balance) of the organism with its environment.

The distinction between thermoregulatory behaviour and thermotropism is ill defined. Thermotropism is defined as the turning or movement of a plant or animal in response to a temperature stimulus. (Greek: θερμη-*therme*-heat; τροπη-*trope*-turn) A plant may exhibit thermotropism but is not considered to be thermoregulating. Some aquatic unicellular organisms move to a preferred ambient temperature; however, whether this is thermotropism or thermoregulatory behaviour may be disputed. For these reasons, the term is usually restricted in thermal physiology to patterns of behaviour controlled by a nervous system.

The complex patterns of somato-motor activities that serve as behavioural thermo-effector responses to heat and cold of temperature regulators in modifying their conditions of heat exchange with the environment, involve a wide variety of performances (e.g. moving to a different thermal ambiance, changes in posture, wetting of body surfaces, change of microclimate by nest building, parental behaviour, huddling) and, in humans, also include voluntary exercise and cultural achievements (clothing, housing, air conditioning, etc.). In endothermic temperature regulators, the reduced demand for autonomic temperature regulation is a potential result of behavioural temperature regulation that, however, competes with other, non-thermal, behavioural drives (e.g. search for food).

Behavioural temperature regulation is characterized as operant, if it is acquired in an experimental condition only after training guided by an experimenter. Otherwise, it is defined as natural, but this may also involve learning. In humans, the distinction between natural and operant behavioural temperature regulation becomes arbitrary in many instances.

Temperature regulation, chemical (obsolete)

Body temperature regulation involving changes in heat production.

This can be due to

1. voluntary muscle movements;
2. involuntary muscle movements (e.g. shivering);
3. non-shivering thermogenesis;
4. increase or decrease in basal metabolic rate (*BMR*).

Temperature regulation, physical (obsolete)

Body temperature regulation involving control of the rate of heat flow into or out of an organism.

The responses involved in such regulation consist of those autonomic and behavioural responses that vary the thermal conductance of peripheral tissues, but not of those behavioural responses which involve alteration of the local environment. For example,

1. changes in peripheral vasomotor tone;
2. piloerection;
3. evaporation of water from skin (following sweating, saliva spreading, wallowing) and from respiratory tract surfaces;
4. changes in body conformation.

Temperature regulation, physiological

1. Both autonomic and behavioural temperature regulation (preferred).
2. Obsolete synonym for autonomic temperature regulation.

(*Continued*)

Table 3.1 (*Continued*)

Traditionally, mammalian thermoregulatory physiology has been concerned with those responses to heat or cold that do not depend on behaviour. These responses are autonomic. Autonomic responses are generally referred to as physiological responses, but behavioural responses are also physiological (Physiology = the science of the normal functions and phenomena of living things, *SOED*). Thus, physiological thermoregulatory responses properly consist of both autonomic and behavioural responses.
Temperature regulator
An organism that regulates its body temperature to some extent by autonomic and/or behavioural processes. Antonym: temperature conformer.
Tachymetabolism is the property of endothermic temperature regulators in which core temperature appears as the controlled variable, with arbitrarily defined degrees of thermostability (homeothermy, heterothermy). Bradymetabolism does not exclude that such species may be ectothermic temperature regulators, although mainly by means of behavioural temperature regulation, because thermo-effectors of autonomic temperature regulation are vestigial or absent.
Temperature conformer
An organism, the core temperature of which varies as a proportional function of ambient temperature; an animal without effective temperature regulation by autonomic or behavioural means. Synonym: poikilothermy. Antonym: temperature regulator.

3.2.1.4.1 *Core temperature*

Deep body temperature was the main interest of early clinical thermometry as a relationship between fever and elevated deep body temperature was established by the mid of the 19th century. The fact that deep body is stable in different climatic conditions that result in various skin temperatures was known for a long time. However, the German physiologist J. Aschoff was one of the first who used consequently the term "core temperature". He based his differentiation between core and shell on the observation that at rest most of metabolic heat is produced in inner organs located inside the chest and abdomen. In other words, core temperature represents the intensity of heat stored in the inner organs. Changes in core temperature are related to increase or decrease of heat content that in turn affects the temperature gradient from the core to the surface.

Figure 3.4. Shows the influence of ambient temperature on the magnitude of insulation of the shell. At ambient temperatures close to core temperature, surface temperatures show little differences throughout the body, but in cold conditions, some regions such as the forehead are still close to core temperature, but others such as regions in the periphery of the extremities may present with temperature differences to the core of more than 20°C.

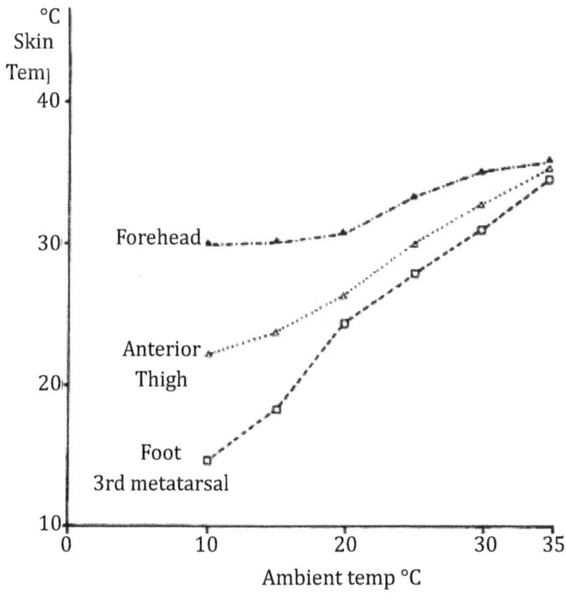

Figure 3.4 Influence of ambient temperature on the skin temperature in three areas of the body, when core temperature is stable.

Due to differences in basic metabolism of distinct deep tissues, their temperature varies slightly. Table 3.2 shows the temperatures obtained at various body sites in a resting subject and their deviation from rectal temperature, 36.85°C.

This compilation from more recent publications shows the offset to pulmonary artery temperature at different sites used for determination and monitoring of core temperatures in patients under intensive care or cardiac surgery respectively.

Table 3.2 Temperatures inside the body core at rest [3]

Site	Low °C (ΔT_r)	High °C (ΔT_r)
Mouth	36.40 (−0.45)	36.55 (−0.30)
Oesophagus	36.55 (−0.30)	36.65 (−0.20)
Stomach	36.65 (−0.20)	36.75 (−0.10)
Liver	36.60 (−0.25)	36.75 (−0.05)
Vagina	36.75 (−0.05)	36.90 (+0.05)
Rectum	36.85	36.85
Brain (mainly hypothalamus)	36.60 (−0.25)	36.90 (+0.05)
Nasopharynx	36.40 (−0.45)	36.45 (−0.40)
Tympanic membrane		
Ambient air at + 10°C	36.45 (−0.40)	36.45 (−0.40)
Ambient air at +40°C	37.25 (+0.40)	37.25 (+0.40)
External auditory meatus	36.35 (−0.50)	36.75 (−0.10)
Urine (at the meatus)	36.70 (−0.15)	36.75 (−0.10)
Cardiovascular system		
Pulmonary artery	36.60 (−0.25)	36.70 (−0.15)
Jugular vein		
Higher part	36.75 (−0.05)	36.75 (−0.05)
Lower part	36.60 (−0.25)	36.65 (−0.20)
Vena cava	36.55 (−0.30)	36.60 (−0.25)
Renal vein	36.60 (−0.25)	36.70 (−0.15)
Coronary sinus	36.75 (−0.05)	36.90 (+0.05)
Right atrium	36.60 (−0.25)	36.70 (−0.15)
Right ventricle	36.65 (−0.20)	36.70 (−0.15)
Left ventricle	36.60 (−0.25)	36.70 (−0.15)

Although the reference values in Tables 3.2 and 3.3 differ, the offset values from the reference agree quite well. The biggest variation of difference to either reference temperature was obtained in ear-based measurements.

Table 3.3 Difference between pulmonary artery temperatures to other body temperatures

Site	Number of measurements (number of patients	Mean±SD or range for pulmonary artery (°C) Mean ΔT_{pa} ±SD (°C)	Reference
Pulmonary artery	659 (41 Pat)	37.42±0.82	Moran et al. [13]
	529 (42 Pat)	33.7 to 40.2	Lefrant et al. [14]
	356 (38 Pat)	34.4 to 38.8	Erickson and Kirklin [15]
	234 (18 Pat)	34.1 to 40.0	Robinson et al. [16]
Bladder	1761(92 Pat)	-0.10±0.12	Moran et al. [13]
	227 (22 Pat)	-0.21±0.20	Lefrant et al. [14]
	318(27 Pat)	0.03±0.23	Erickson and Kirklin [15]
Rectum	529 (42 Pat)	-0.07±0.40	Lefrant et al. [14]
	234 (18 Pat)	-0.4±1.0	Robinson et al. [16]
Oesophagus	529 (42 Pat)	0.11±0.30	Lefrant et al. [14]
	234 (18 Pat)	0.0±0.5	Robinson et al. [16]
Oral	288 (37 Pat)	0.05±0.26°	Erickson and Kirklin [15]
Ear-based	648 (110? Pat)	-0.52±0.15	Moran et al. [13]
	347 (37 Pat)	0.07±0.41	Erickson and Kirklin [15]
	234 (18 Pat)	-0.4±0.5	Robinson et al. [16]
Axillary	634 (110? Pat)	-0.38±0.04	Moran et al. [13]
	529 (42 Pat)	0.27±0.45	Lefrant et al. [14]
	356 (27 Pat)	-0.68±0.57	Erickson and Kirklin [15]
	228 (18 Pat)	0.2±1.0	Robinson et al. [16]
Inguinal	529 (42 Pat)	0.17±0.48	Lefrant et al. [14]

Intracranial temperature

Progress in neurosurgery and in technology of thermometric devices permits now in vivo temperature measurements of intracranial structures including the brain in humans. The course of intraventricular, epidural and rectal temperature was compared in 28 intensive care patients with consciousness levels between 1 and 6 as measured by the Reaction Level Scale 85 [17]. There was a tendency that intraventricular and rectal temperature increased with decreasing consciousness. The epidural temperature was always lower than the intraventricular temperature (offset 0.42°C to 0.62°C), whilst the difference between intraventricular and rectal temperature varied between 0.27°C and 0.40°C. This study suggests that temperature recordings taken near the brain surface may not adequately reflect brain temperatures.

In 31 Japanese patients with severe subarachnoid haemorrhage, the relationship between intraventricular and bladder temperature predicted a fatal outcome. Patients with brain temperature lower than extra-cerebral core temperature at any time during the course died of acute brain swelling. In contrast, patients with higher brain than core temperature throughout the course survived [18].

A systematic review based on 15 studies which compared brain temperatures (intraventricular, intra-parenchymal, epidural) with any other core temperature (rectal, bladder, pulmonary artery, oesophageal, tympanic) reported higher brain than extra-cerebral core temperatures in most participants [19]. Offsets between the brain and other core temperatures are provided in Table 3.4.

Table 3.4 Differences between brain temperatures and other core temperatures

Comparison	Range of offset	Number of studies
Rectal Temperature Versus Brain Temperature	–0.3 to –2.0°C	13
Bladder Temperature Versus Brain Temperature	–0.5 to –2.5°C	5
Pulmonary Artery Temperature Versus Brain Temperature	–0.3±0.3°C (range: 0.7 to –2.3°C)	1

Bladder and urine temperature

Bladder temperature is measured with insertion of a thermosensor by use of a catheter. It is an alternative to rectal temperature monitoring in intensive care. Good agreement between bladder and pulmonary artery temperature or rectal temperature has been shown [20]. Urine temperature was suggested as an estimation for core temperature in industrial workers in hot or cold environments [21].

Temperatures in the digestive tract

Core temperature was measured in many sites of the digestive tract including the oral cave, the oesophagus, stomach, small and large intestine and rectum. Mouth and rectum are traditional sites for fever measurement, oesophagus and rectum are used for core temperature monitoring in critically ill patients. Gastrointestinal temperatures, rectal and oesophageal temperatures are often measured in studies of thermal physiology related to physical workload, particularly in sports and exercise.

Oral temperature

Oral or, more exactly, sublingual temperature can only be measured reliably if the cavity is well closed. Based on the pooled data from 20 studies of high to good quality, reference values for healthy subjects were reported with 36.7±1.0°C for men (n = 934) and 36.2±3.0 for women (n = 166). Another systematic review related oral to rectal temperature found under resting conditions the temperature in the mouth lower by 0.50±0.31°C [23]. Under transient conditions such as exercise or large environmental change, difference between rectal and oral temperature was 0.58±0.75°C in favour of the rectal measurement.

Oesophagus

The measurement of oesophageal temperature is carried out by the subject swallowing a temperature probe. It is used in patient monitoring during intensive care or surgery but is also applied in exercise physiology. Under resting conditions and after exercise oesophageal temperatures were always found to be lower than rectal as shown in Table 3.5.

The relationship between the oesophagal probe temperature and intestinal temperature recorded with thermosensitive pills was investigated and limits of agreement between −0.45°C and 0.41°C or −0.40°C and 0.34°C have been reported.

Table 3.5 Comparison of oesophageal (T_{oe}) and rectal temperatures (T_r) during upper and lower body exercise [24]

Condition of measurement	Number (years)	Site	Mean±SE	Reference
Upper body exercise	5 males (25±1.5)	T_{oe} at rest	36.8±0.2	
		T_{oe} at end of exercise	37.3±0.2	
		T_r at rest	37.2±0.2	
		T_r at end of exercise	37.5±0.2	
Lower body exercise (cycling)	6 males (24.5±0.3)	T_{oe} at rest	36.8±0.1	Gass and Gass
		T_{oe} at end of exercise	38.0±0.1	
		T_r at rest	37.3±0.1	
		T_r at end of exercise	38.3±0.1	
Lower body exercise (treadmill)	4 males (27.5±0.9)	T_{oe} at rest	36.9±0.2	
		T_{oe} at end of exercise	38.2±0.1	
		T_r at rest	37.3±0.1	
		T_r at end of exercise	38.4±0.1	

Intestinal temperature

Ingestible telemetric temperature sensors represent an alternative to oesophageal and rectal temperatures as an index of core temperature during exercise. The analysis based on a limited dataset suggests that the agreement between $T_{oesophageal}$ and $T_{intestinal}$ is as good as, if not better than, the agreement between $T_{oesophageal}$ and T_{rectal} with a systematic bias <0.1°C and 95% limits of agreement within ±0.4°C [25].

Rectal temperature

Rectal temperature is most often measured in clinical medicine or in experimental studies as representing the internal temperature of the body. To obtain a true rectal temperature, the probe must be inserted to a depth of at least 8 cm. When conditions are correct, and the subject is at rest, rectal temperature is one of the highest in the body. Normal reference values based on a large sample of measurements are not yet available. For a small sample of 57 men, a value of 37.0±0.3°C was reported and 37.0±0.2°C for 37 women [22].

Axillary Temperature

The under arm axilla is a convenient site for estimating core temperature in both adults and infants. In elderly healthy adults, the mean of a sample 530 subjects the temperature was 36.0±1.1°C [22]. In children and young people, the temperature measured at the axilla does not agree sufficiently with temperature measured at the rectum to be relied on in clinical situations where accurate measurement is important. In older children or adolescents, the pooled temperature difference between axillary and rectal measurements was 0.92°C with 95% confidence interval of –0.15 to 1.98°C, but the difference in the newborn was only 0.17°C. The variability in results was related to the age of the child and duration of placement time of the measuring device [26]. In the newborn, an excellent agreement was found between axillary and rectal temperature measurements [27].

Ear-based temperature

This area, being easy to explore, has long been a popular site for body temperature measurement. The temperature of the cavity is, however, particularly dependent on ambient conditions, and therefore this temperature may not represent a true deep temperature. Tympanic temperature can be measured directly and indirectly. Direct measurements can be obtained by holding a thermocouple against the tympanic membrane. The major disadvantage of this method is the risk of pain or damage. The measurement can be confounded easily by insufficient isolation of the probe and slight displacement of the probe from the tympanum.

Tympanic temperature is now determined indirectly with an Infrared (IR) thermometer. The IR tympanic thermometer measures the thermal radiation emitted by the tympanic membrane. This method increases safety, speed and comfort of tympanic measurements.

Normal values of tympanic membrane radiometry based on measurements in 564 men (36.5±1.0°C) and 861 women

(36.6±0.9°C) have been reported [22]. In children, the pooled mean temperature difference (rectal minus infrared ear thermometry) was 0·29°C (95% limits of agreement −0·74 to 1·32). Although these differences are small, the wide limits of agreement mean that ear temperature is not a good approximation of rectal temperature [28].

There is a major risk of underestimating the real tympanic temperature and thus incorrectly estimating core temperature. The ear canal is shaped irregularly, and a measurement is easily taken too superficial or aiming in the wrong direction [29].

Temporal artery

Another technique for estimating core temperature is based on measurement of infrared radiation emitted from the region of the superficial temporal artery. As few validation trials have been carried out so far, reliability remains questionable. In a study published in 2014, the median temperature in 294 children was 37.1°C (range: 35.6–39.8°C) with mean differences to rectal or tympanic temperature of 0°C and 0.49°C respectively. However, the limits of agreement were between −1.31°C and 1.33°C for rectal temperature, and −0.71°C and 1.69°C for tympanic temperature [30].

Infrared imaging inner canthus of the eye

This method of mapping surface temperatures was first applied for mass fever screening in pandemic alert. The inadequate use of this technique led to the requirement of an international standard, which was published in 2008/2009 [31]. A review paper published in 2009, showed that fever screening based on infrared radiometry showed both high diagnostic sensitivity and high specificity when the target area was the inner canthus of the eye [32].

The International Standard states *"The current evidence indicates that the region medially adjacent to the inner canthi is the preferred site for fever screening due to the stability of that measurement. This is because this region is directly over the inner carotid artery"* The ISO standard has been updated and extended in 2017 [33].

The validity of the technique for estimating core temperature should be further investigated. In a study that followed the full criteria set out by the ISO standard, there was good agreement in more than 400 children between inner canthi and axillary temperatures [34]. Less agreement with other core temperatures such as oesophagus or rectum have been reported. These were children all examined in a specialist hospital and heir diagnosis as febrile or non-febrile was established but blinded from the investigators.

There are several important aspects to the ISO protocol that included the calibration, resolution and proven stability of the IR camera, the range of ambient temperature, and the precise location of the regions of interest over the inner canthi taken from a full-frontal face thermogram. Only when all these factors are strictly controlled, can the procedure be reliably employed for fever screening.

Factors causing change of core temperature

Deep body temperature is never constant except in a steady state, which can only be obtained in experimental conditions. Physiologically, deep temperature is always changing to some extent and is dependent on many factors.

The first important factor is *age.* Differences may exist in the circadian rhythm of thermoregulation between young and older adults. In addition, core temperature is subject to *cyclical* changes, e.g. circadian, hormonal in females.

Circadian Rhythms

The existence of a circadian variation in body temperature has been well known since the 19th century. Temperature increases during the day, reaching its maximal level in the evening. The minimal value is observed at the end of the night (Fig. 3.5). Several types of circadian variations have been described. They are mainly dependent on the time of the maximal peak (*acrophase*), which can vary from one subject to another.

A complex relationship exists between central temperature variations and waking-sleeping alternation, i.e. light-darkness and physical activity variation. The period sequence for central temperature can be considered as a specific property of the body, which can be modified by an external synchronization.

Temperature Variation during the Menstrual Cycle

Central temperature is well known to be a cyclical variant in the female. The mean temperature of the circadian cycle is lower during the postmenstrual period. It increases immediately after ovulation and is then maintained at a higher level through the premenstrual time. The rise is around 0.5°C but can vary in amplitude from one subject to another.

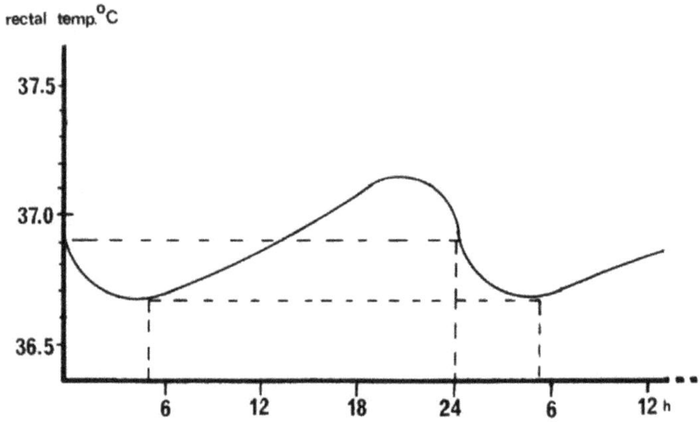

Figure 3.5 Diurnal variation of rectal temperature.

From a practical point of view, the menstrual change in deep body temperature is only obvious if the measurement is taken under the same conditions each day, to avoid interference by the circadian pattern.

3.2.1.4.2 *Shell temperature*

The temperature distribution of the shell of which the skin is part, depends on the thermal balance. As the skin is the interface between the body and the environment, its temperature is influenced by both internal and external conditions. A constant gradient from the core to the skin creates a steady state in distribution in all layers of the shell temperature, which becomes a function of the environmental conditions, when external factors change.

A subject is in his thermoneutral zone when temperature regulation is achieved only by control of sensible heat loss, i.e. without regulatory changes in metabolic heat production or evaporative heat loss. Body heat loss and basal body heat production are then equal. If the subject is at rest and unclothed, this state occurs in an ambient temperature of 30°C with low humidity and air movement of less than 0.5 ms^{-1}.

Physiological research journals reported skin surface temperatures that have been almost exclusively obtained with surface skin probes. Each probe can only measure at one small location that cannot represent the variable distribution in surface temperature. This problem is easily overcome by infrared thermal imaging, a technique that emerged during the 1960s. Remote sensing provided the opportunity to record simultaneously skin temperature in various anatomical regions. Since the 1980s, skin temperature of healthy subjects has been mapped and published.

Table 3.6 shows regional skin temperature values recorded by infrared thermal imaging. In each gender group, the mean values of the right and left side were calculated. In the first two studies, the average temperature is based on values from men and women.

Table 3.6 Regional skin temperatures

Study		Chudecka and Lubkowska (2015) [35]	Marins et al., (2014) [36]	Marins et al., (2014) [37]	
Gender		100 males 100 females	18 males 26 females	103 males	117 females
Region of interest	Dorsal arm	31.1±0.3°C	29.0±.5°C	29.4±1.2°C	28.5±1.4°C
	Anterior arm	33.3±0.3°C	30.8±1.2°C	31.3±1.3°C	30.9±1.3°C
	Dorsal forearm	320.3±0.3°C	30.2±0.8°C	30.7±1.1°C	30.1±1.3°C
	Anterior forearm		30.6±1.0°C	30.9±1.2°C	30.5±1.2°C
	Dorsal hand	31.8±0.3°C	28.6±1.6°C	28.0±2.1°C	28.3±1.8°C
	Palmar hand	31.9±0.3°C	29.0±1.7°C	28.3±2.6°C	28.5±2.2°C
	Dorsal thigh	30.8±0.3°C	29.6±0.9°C	30.3±1.1°C	29.2±1.1°C
	Anterior thigh	31.6±0.3°C	29.5±0.9°C	29.7±1.2°C	28.7±1.3°C
	Dorsal lower legs	31.2±0.3°C	29.6±0.9°C	29.0±1.3°C	29.2±1.1°C
	Anterior lower legs	31.6±0.2°C	30.0±1.0°C	30.2±1.4°C	30.1±1.2°C
Ambient temperature		25°C	19±0.3°C	21±0.3°C	

Table 3.6 shows the effect of ambient temperatures on skin temperatures after 20 min of acclimation. Values at 25°C are higher than those obtained in the study conducted at 19°C. Overall differences between males and females were small, but significant statistical differences are highlighted in the table.

Thermal imaging has confirmed earlier concepts of similar temperature distribution in corresponding anatomical regions of the body. A close agreement of mean temperature in contra lateral measurement areas has been termed thermal symmetry. A side-to-side difference in mean temperature of 1.0°C was regarded as abnormal. It was suggested that the cut-off value was based on mean + 2 standard deviations. Vardasca provided a comparison of in side-to-side differences in skin temperatures of the extremities as shown in four previous publications (Table 3.7). In a recent study performed in 39 subjects with modern equipment, smaller mean differences in corresponding temperatures were reported [38].

Table 3.7 Reference values of side-to-side differences in skin temperatures

Study		**Uematsu (1985) [39]**	**Goodman (1987) [40]**	**Uematsu (1988) [41]**	**Niu (2001) [42]**
Region of interest	Dorsal arm	0.2±0.2°C	NODATA	0.4±0.3°C	0.5±0.4°C
	Anterior arm	0.1±0.2°C	NODATA	0.3±0.23°C	0.5±0.4°C
	Dorsal forearm	0.3±0.2°C	1.0±0.10°C	0.3±0.2°C	0.5±0.3°C
	Anterior forearm	0.2±0.2°C	0.4±0.1°C	0.3±0.3°C	0.3±0.2°C
	Dorsal hand	0.4±0.1°C	0.6±0.1°C	0.3±0.3°C	0.4±0.3°C
	Palmar hand	NODATA	0.6±0.1°C	0.3±0.2°C	0.4±0.3°C
	Dorsal thigh	0.2±0.1°C	0.4±0.1°C	0.2±0.2°C	0.2±0.1°C
	Anterior thigh	0.1±0.1°C	0.4±0.1°C	0.2±0.2°C	0.2±0.2°C
	Dorsal lower legs	NODATA	0.4±0.03°C	0.3±0.2°C	0.2±0.2°C
	Anterior legs	0.3±0.3°C	0.4±0.04°C	0.3±0.2°C	0.3±0.3°C
	Dorsal feet	0.3±0.2°C	NODATA	0.4±0.3°C	0.4±0.3°C
	Plantar feet	NODATA	NODATA	0.4±0.3°C	0.4±0.3°C

The mean temperatures of finger joints recorded by infrared thermography in the posterior view from 42 subjects who did not present any arthritic symptoms [43] are shown in Fig. 3.6. The sample included 32 women and 19 men with a mean age of 47.7±16.9 years.

A systematic review collected data from studies reporting knee temperatures from subjects with various health conditions and obtained with different methods. In healthy subjects, the mean temperatures varied with respect to the applied measurement technique and the site of measurement. Mean intra-articular knee temperature was 32.6±0.9°C (95% confidence interval: 31.5 to 33.7°C). Based on contact temperature measurements, the mean temperature of the anterior knee was 30.5±1.1°C. Determined with radiometers, the corresponding mean temperature was 28.7±0.6°C (95% confidence interval: 27.9 to 29.5). In thermal images from healthy subjects, the mean temperature of the anterior knee was 29.5±1.6°C (95% confidence interval: 28.5 to 30.5°C) and 33.1±0.0°C of the posterior knee. The side difference of temperature was 0.2±1°C for both the anterior and the posterior knee [44].

Mean skin temperature

The "Glossary of Terms for Thermal Physiology" has defined, **mean skin temperature (T_{sk})** as the sum of the products of the area of each regional surface element (A_i) and its mean temperature (T_i) divided by the **total body**(surface) **area.**

$$T_{sk} = \Sigma(A_i \cdot T_i)/A_b \ [°C]$$

Mean skin temperature can be used as a physical variable in the calculation of heat balance or of heat content of the body. It is, however, not necessarily a good estimate of what might be sensed as a **mean skin temperature** according to the integrated neural input of the cutaneous **thermo receptors,** because different surface regions may differ in their importance as **thermo sensor** sites.

Mean skin temperature must not be confused with the mean temperature of a local skin area of an anatomical region. In thermal image analysis, mean temperature is a common expression relating to a specific region of interest.

Methods for determining mean skin temperature will be discussed in Section 3.3.1.2.

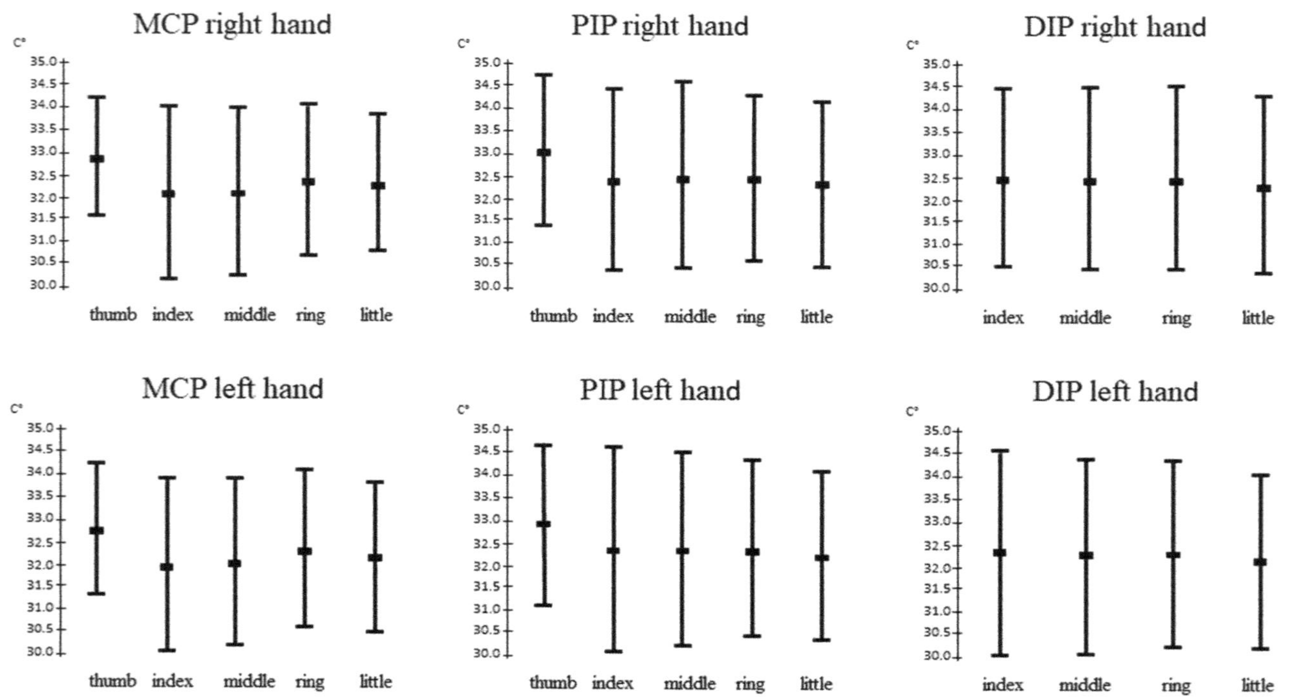

Figure 3.6 Mean±standard deviation of temperature of the metacarpophalangeal (MCP), proximal (PIP) and distal interphalangeal (DIP) joints of healthy subjects.

3.3 Heat Exchange with the Environment

For the maintenance of the heat content of an object exposed to external heat load or heat loss, there are two possible mechanisms. One is to modify the heat transfer to the environment by changing the thermal resistance and/or heat dissipation, the other is increasing or decreasing heat production. For both solutions, a regulation loop is necessary, constructed of sensors, a processor unit and effectors.

As physiological thermoregulation aims to keep the temperature of inner organs within a narrow range, thermoregulation results at environmental temperatures below the core temperature in a constant temperature gradient directed from deep body tissues to the body's surface. Therefore, the magnitude of the gradient depends primarily on the difference between deep body and environmental temperature.

It must be noted that the heat content of deep tissues varies in both intensity, i.e. temperature, and shape of distribution in the three-dimensional space, and that heat travels by either conduction or convection to the surface. In absence of contact between the skin and other surfaces and undisturbed natural convection around the body, infrared radiation is the main mechanism of heat transfer to the air environment.

3.3.1 Heat Balance Equation

A central role in quantitative thermal physiology is taken by the **Body heat balance equation.** Its definition in the "Glossary of Terms for Thermal Physiology" is based on heat flows, which is the net rate at which a subject generates and exchanges heat with its environment (Table 3.8). Although this equation is intended to understand the physical principal that results in stable core temperature, this measure of heat intensity is not considered.

In more simple terms, the body heat equation describes the thermal *steady state of a* living body in which both superficial and deep temperatures are constant, but not equal. This implies that the physical removal of heat by the ambiance exactly compensates for the heat produced by the body. Generally, a steady state is achieved when heat loss is equal to heat gain. If the sign (+) is conventionally attributed to the gains and the sign (–) to the losses, their algebraic sum will be equal to zero: (gains) – (losses) = 0.

Table 3.8 Terminology related to the body heat equation [2]

Body heat balance equation (actually: **heat flow balance equation**)
This is a mathematical expression that describes the net rate at which a subject generates and exchanges heat with its environment. The dimension of the equation and its terms respectively, are in watts [W] thereby constituting **heat flows**, but often are also expressed in relation to unit area of body surface [$W \cdot m^{-2}$], to unit body mass [$W \cdot kg^{-1}$], or to unit body volume [$W \cdot m^{-3}$].

$$S = M - (W) - (E) - (C) - (K) - (R).$$

S = **storage of body heat** (positive = increase in **body heat content**; negative = decrease in **body heat content**).
M = **metabolic energy transformation** (always positive in a living organism) = **metabolic rate.**
W = **work rate, positive** (= **useful mechanical power accomplished**; negative = mechanical power absorbed by body: **work rate, negative**).
E = **evaporative heat transfer** (positive = **evaporative heat loss**; negative = **evaporative heat gain**).
C = **convective heat transfer** (positive = transfer to the environment; negative = transfer into the body).
K= **conductive heat transfer** (positive = transfer to the environment; negative = transfer into the body).
R = **radiant heat exchange** (positive = heat transfer to the environment; negative = heat absorption by the body).

The **body heat balance equation** is designed to describe the prevailing physiological situation of a **homeothermic** (e.g. human) subject. At a given **metabolic energy transformation**, the subject transfers some mechanical energy to its ambiance (→**work rate, positive**), and, if the ambient thermal conditions permit it, heat from the body to the environment by evaporation, conduction, convection, and radiation, which eventually results in the stabilization of **body heat content** with storage S = 0. However, there are physiological conditions in which the body absorbs mechanical energy, ultimately as heat (→**work rate, negative**), and/or one or the other of the terms describing the exchange of **heat flow** with the environment may become negative.

In a **core/shell** treatment of the body, there are two **heat balance equations**:

$$S_{core} = M - (W) - (H) - REHL$$
$$S_{shell} = H - (E) - (C) - (R) - (K)$$

with *REHL* = **respiratory evaporative heat loss (*REHL*)** and *H* = **heat transfer** (conductive and convective heat flow) between **core** and **shell**. The latter is proportional to the difference of mean **core** and **mean skin temperature** with the proportionality constant → **conductance k.**

Heat content, body

The product of the body mass, its average **specific heat**, and the absolute **mean body temperature**. [J]

The actual value of this term is seldom calculated. It is used only in the determination of **heat storage→storage of body heat.**

Storage of body heat (*S*)

The rate of increase (+) or decrease (–) in the **heat content** [Ws = J] of the body caused by an imbalance between heat production (**metabolic heat transformation**) and **heat loss**, usually expressed in terms of unit area of total body surface (**→Area, total body**). The quantity ***S*** in the **body heat balance equation**. [W], [W $\cdot m^{-2}$], [W$\cdot$ m^{-3}] or [W $\cdot$ kg^{-1}]

Total heat production

The rate of transformation of chemical energy into heat in an organism (**metabolic heat production ($\dot{H}$)**) plus any heat flow liberated within the body resulting from work done on the organism by an external force (**negative work rate**). [W], [W$\cdot m^{-2}$], [W$\cdot m^{-3}$], [W$\cdot kg^{-1}$] [3].

During **positive work** and when no work is being done on or by the organism, **total heat production** equals **metabolic heat production ($\dot{H}$)** [3].

Area, total body (A_b)

The area of the outer surface of a body, assumed smooth. [m^2] [3].

Direct measurements of surface area are difficult and subject to appreciable error. Surface area is usually estimated from a formula such as that of Meeh (1879) that relates **total body area (A_b)** to body weight (*W*): $A_b = k \cdot W^{2/3}$. This is a particular case of Huxley's allometric law. Estimates of the value of *k* vary widely between and within species but are generally between 0.07 and 0.11 when A_b is in m^2 and *W* is in kg (Spector, 1956). The DuBois formula (**→Area, DuBois**), used in the estimation of the **total body area**, relates **total body area** to both mass and height. Estimates of **total body area** made with either Meeh's formula or the DuBois formula has limited accuracy, and reference should be made to the original direct measurements from which **total body area** was calculated. References: Huxley JS, *Problems of Relative Growth,* London: Methuen, 1932; Meeh K, Z. *Biol.* 1879, 15, 425; Spector WS, *Handbook of Biological Data*, Philadelphia, P.A.: Saunders, 1956 [3].

(*Continued*)

Table 3.8 *(Continued)*

Area, DuBois (A_D) The total surface area in square meters of a nude human as estimated by the formula of DuBois based on the height, H [m], and mass, W [kg]. Reference: DuBois EF, *Basal Metabolism in Health and Disease*, Philadelphia, PA, Lea & Febiger, 1927. $A_D = 0.202 \cdot W^{0.425} \cdot H^{0.725}$ [m²] [3].
Work rate, negative (−W) The **rate** of work **(→power)** done on an organism by an external force. The quantity (−W) in the **body heat balance equation**. [[W], $W \cdot m^{-2}$], [$W \cdot m^{-3}$], or [$W \cdot kg^{-1}$] Antonym: **work rate, positive** [3].
Work rate, positive (+W) The rate of work **(→power)** done by an organism on an external system. The quantity (+W) in the **body heat balance equation**. [W], [$W \cdot m^{-2}$], [$W \cdot m^{-3}$], or [$W \cdot kg^{-1}$] Synonym: mechanical energy, **work production, useful work accomplished**. Antonym: **work rate, negative** [3].
Temperature, mean body (T_b) (1) Ideally, the sum of the products of the **heat capacities** ($c_i \cdot m_i$) and temperature (T_i) of all tissues of the body divided by the total heat capacity of the organism: $\bar{T}_b = \Sigma(c_i \cdot m_i \cdot T_i)/\Sigma(c_i \cdot m_i)$. (2) The total temperature signal generated by all **thermosensors** distributed in the **core** and **shell**. (2) should be different from (1). For discussion, see Minard D, in: *Physiological and Behavioral Temperature Regulation*, ed. Hardy JD, Gagge AP, Stolwijk JAJ, Springfield, I11., Thomas, 1970, p. 345. **Mean body temperature** often is estimated approximately from measurements of a representative **core temperature** (T_c) and **mean skin temperature** (T_{sk}) according to $T_b = a_1 \cdot T_c + a_2 \cdot \bar{T}_{sk}$, with $a_1 + a_2 = 1$, in which the factors a_1 and a_2 represent the empirically determined contributions of **thermal core and shell** to **mean body temperature**.
Temperature, mean skin (T_{sk}) The sum of the products of the area of each regional surface element (A_i) and its mean temperature (T_i) divided by the **total body** (surface) **area.** $\bar{T}_{sk} = \Sigma(A_i \cdot T_i)/A_b$ [°C]

Note: **Mean skin temperature** can be used as a physical variable in the calculation of heat balance or of heat content of the body. It is, however, not necessarily a good estimate of what might be sensed as a **mean skin temperature** according to the integrated neural input of the cutaneous **thermo receptors,** because different surface regions may differ in their importance as **thermo sensor** sites.

Heat transfer from the core to the surface can be described in analogy to a two-step waterfall.

The subject produces a certain amount of metabolic heat, *M*, per unit of time (analogous to a certain mass of water per unit of time). Just as water falls from one level to another, so heat is transferred from the core (at T_d) to the skin (at T_s) (h_b is the transfer coefficient), owing to the temperature gradient which is proportional to the difference of mean core and mean skin temperature with the proportionality constant.

In a steady state, this amount of heat per unit of time is further transferred from the skin to the environment (at T_a), also owing to the temperature gradient. In the example shown in Fig. 3.7., the temperature gradient is slightly higher than that existing between core and skin, but the transfer coefficient *h* is lower. However, the areas of the three rectangles are the same, as they represent the same amount of heat transferred per unit of time.

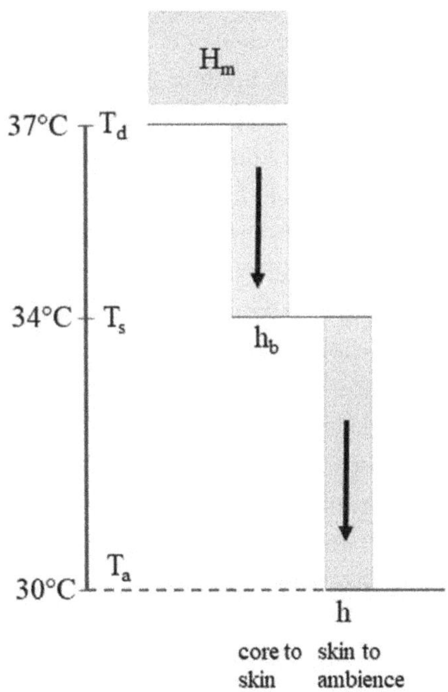

Figure 3.7 Waterfall analogy of heat transfer.

3.3.2 Determination of Mean Skin Temperature

Mean skin temperature was computed from measurements in various body sites. The number of proposed measurement sites varied between 3 and 15 [3]. One of these formulas for estimation of mean skin temperature was included in the ISO Standard for mean skin temperature calculation based on contact measurement on the left dorsal neck, the right scapular region, the dorsum of the left hand adjacent to the third metacarpophalangeal joint and the right upper tibia [45]. A recent study investigated the distribution of skin temperature at the proposed measurement sites by thermal imaging and found in all investigated location high homogeneity and normal distribution of skin temperature [46].

Thermal imaging is seldom used for determining mean body skin temperature. In a case study that compared the ISO method with total body thermal images found the estimated mean skin temperature was approximately 2°C higher than the mean temperature readings obtained from the total body view [47].

The heat flow from living body is clearly related to the surface area and this relationship is relevant for all forms of heat exchange, i.e. conduction, convection, radiation and evaporative heat loss.

Body posture does change the proportion of areas of which heat is transferred by conduction, convection or radiation. If a subject is in the supine position, he will exchange heat by conduction through the supporting surface, and by radiation from the exposed surface. Dependent on the airflow, convection contributes to heat exchange on the exposed surface. Posture affects also the size of the exposed skin surface. In cold exposure, a typical behavioural response is to reduce surface area in a squatting position. Table 3.9 shows estimated ratio of effective radiating surface area of the body to total body area in different positions.

Posture affects also natural convective boundary layer flow as shown in Fig. 3.8 in still air. When lying, the flows are generally slower than when standing. Sitting produces flows that are intermediate between lying and standing.

Table 3.9 Relation between posture and effective radiating surface area

Position	Radiating surface/ total surface (A_r/A)	Authors[a]
	0.72	Bedford (1935)
	0.66	Guibert and Taylor (1952)
	0.70	Guibert and Taylor (1952)
	0.70	Fanger (1970)
	0.82	Bohnenkamp and Ernst (1931)
	0.82	Bedford (1935)
	0.78	Hardy and Dubois (1938)
	0.78	Guibert and Taylor (1952)
	0.84	Colin et al. (1966)

[a]Cited in Colin et al. (1970).

Figure 3.8 Convective natural convective boundary layer flow in different postures.

3.3.3 Clothing

The thermo neutral zone was defined earlier as the range of ambient temperature at which temperature regulation is achieved only by control of sensible heat loss. Figure 3.9 shows heat transfer under conditions of thermal neutrality for a man, either nude or clothed. In both cases, man produces the same amount of heat in each unit of time.

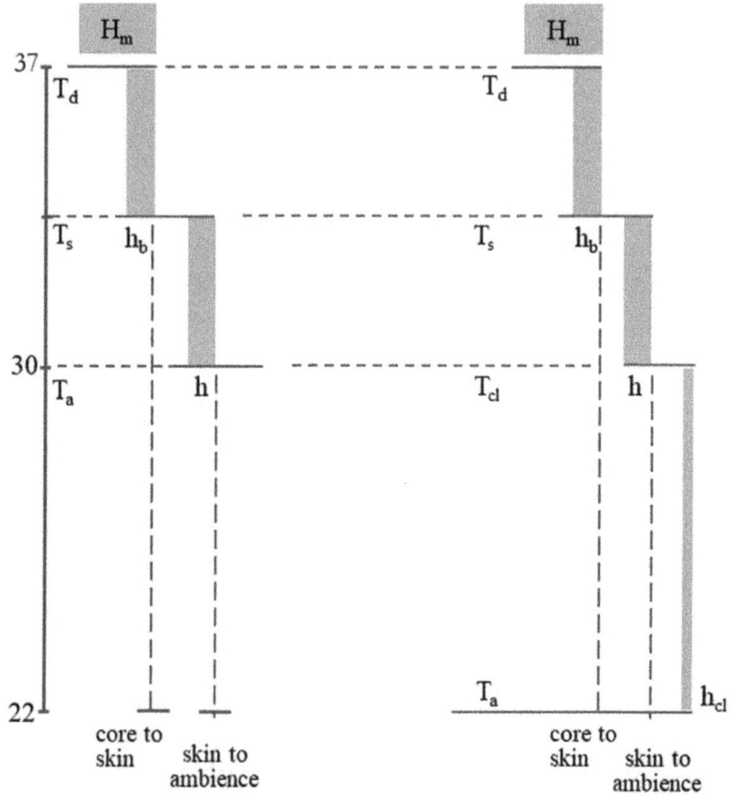

Figure 3.9 Heat transfer in nude or clothed man.

Metabolic heat (M) flows from core (at T_d) to periphery (at T_s) and from periphery to environment (at T_a). When the subject is nude, the environment must be at 30°C to ensure the full removal of M. When the subject is clothed, a further stage for heat transfer is introduced: transfer from the immediate environment of the skin (created by clothing). M must be removed from this local environment by transfer through the clothing. Therefore, the thermal impedance of the clothing will govern the final temperature. In the example shown the final temperature is 22°C.

Clothing resists heat and moisture transfer between skin and environment. This can protect against extreme heat and cold, but also hampers heat loss during physical effort.

Clothing insulation (m^2°CW^{-1} or in clo; 1 clo = 0.155 m^2°CW^{-1}) is expressed either as total insulation (I_t, includes surface air layer) or so-called "intrinsic insulation" (I_{cl}, clothing with enclosed air layers only).Values for clothing insulation (I_T for clothing including surface air layer, or I_{cl} for clothing only) and vapour resistance (*Re*) can be measured, but usually these are estimated from lists of insulations for numerous clothing ensembles [48].

Cosmetics may influence infrared emittance from the skin and change infrared-based temperature readings.

3.3.4 Ambient Conditions

The nature of the ambient medium affects heat transfer from the human body. The physical principles are described in Chapter 1.

Thermal properties such as heat capacity, thermal conductivity, heat transfer rate and heat generation rate of selected tissues are provided in Table 3.10. Heat transfer rate for the three main layers of the shell temperature, i.e. skin, fat and muscle, are in the ratio 3 to 1 to 1. As the blood flow of the skin can vary between 75 ml/min in a state of cold-induced vasoconstriction and 8 l/min under heat-induced vasodilation, the thermal resistance of the perfused tissue will change accordingly.

The following example for conductive heat transfer is taken from Houdas [3].

In a transient state (see Chapter 1), these characteristics are the thermal conductivity k, the density ρ, and the specific heat c. These three parameters are generally linked in the product $k\rho c$. The greater the product, the higher the capacities for heat transfer. (Table 3.11 shows the values of $k\rho c$ of some common materials.) In practice, this explains why the sensation of the temperature of a given object depends not only on its actual temperature, but even more on its $k\rho c$ values. An iron object at 0°C gives a sensation of cold more than an identical object made of wood at the same temperature, because the iron removes more heat from the skin than wood. Conversely, if the object is at a higher temperature than the skin, the greater the $k\rho c$, the more heat is transmitted to the skin, and, therefore, the greater the sensation of warmth that is experienced.

Table 3.10 Thermal properties of selected tissues in thermoneutral conditions [49, 50]

Tissue	Heat capacity (J/kg/°C)	Thermal conductivity (W/m/°C)	Heat transfer rate (ml/min/kg)	Heat generation rate (W/kg)
Skin	3390±233	0.37±0.06	106±37	1.65±0.57
White Fat	2348±372	0.21±0.02	33±13	0.51±0.2
Brown Fat				~300 [49]
Muscle	3421±460	0.49±0.04	39±13	0.96±0.32
Brain	3630±74	0.51±0.02	559±99	11.37±2.01
Skull	1313±295	0.32±0.03	10±0	0.15±0
Breast gland	2960±0	0.33±0.02	150±0	2.32±0
Thyroid Gland	3609±0	0.52±0.02	5624±1403	87.1±21.73
Water	4178	0.60	0	0
Air (−50 to 40°C)	1005	0.020 to 0.027	0	0

Table 3.11 Values of thermal inertia of and sensation of heat or cold given by various substances

Material	Thermal inertia ($J^2 \cdot cm^{-4} \cdot °C^{-2} \cdot s^{-1}$)	Temperature (°C)		
		Threshold of pain (cold)	Range of comfort	Threshold of pain (warmth)
Steel	1.57×10^{-2}	14	29–32	45
Concrete	3×10^{-2}	4	27–34	54
Rubber	3.8×10^{-3}	−12	24–35	67
Wood				
Oak	2.3×10^{-3}	−20	22–35	74
Pine	7.7×10^{-4}	−53	17–39	84
Cork	1.9×10^{-4}	−140	5–42	150

The body exchanges heat by convection with air or sometimes with water, when immersed. These exchanges arc related to the temperature difference between the skin and the air, $T_s - T_a$, and to the exchanging surface area of the body, $A_{c.}$ In general, heat conductivity of water is 20-fold of the conductivity in air that varies between 0.020 and 0.027 W/m/°C in the range of –50 to 40°C (Table 3.11).

3.3.4.1 Inhomogeneous heat distribution

The thermal energy in the ambience of a living body is not necessarily in the state of full equilibration where heat has the same intensities in its volume and heat accumulation with higher temperature does not exist. Contact with the ground or supporting surfaces, touching or gripping tools are common situations in which heat exchange by conduction occurs in daily life.

Homeothermic animals including humans have a warm core zone of temperature covered by a shell zone that varies in extension and insulation capacity. Within the core, small differences in temperature exist and the question arises, whether focal heat accumulations in deep tissue layers can become visible as hot spots in the skin. Such a mechanism was proposed as the rationale for the detection of physiological and pathophysiological processes by infrared thermal imaging including breast cancer [51], inflammation of large blood vessels [52] or activated brown fat in adult humans [53]. Conduction of thermal energy from deep tissues to the body surface is without debate. However, the path of heat conduction remains unexplained and the point of arrival is also unknown. In any volume, heat transfer occurs along temperature gradients and the direction of heat transfer is modified by the geometry of conductive tissue properties. The current knowledge on the geometric distribution of tissue properties such as heat capacity and thermal conductivity does not explain that a small deposit of heat generating tissue is detectable as an area of increased temperature at the surface directly above the heat source.

Turbulent flow within the air or water increases the heat exchange between the living body and the ambient medium. Wind removes heat depending of its velocity and the draught of the air conditioning system directed to the surface of an individual whose thermal image is captured, leads to erroneous temperature readings from a thermal image [54].

Convective cooling with a fan removes the natural convective boundary layer from the skin and disturbs the otherwise uniform distribution of skin temperature. After stopping forced air convection, small hot spots become visible representing perforating vessels that transport warm blood from the muscles to the skin [55]. Those hot spots fade when the natural convective boundary layer is recovering.

Infrared radiation from objects in the near or far environment affects also the heat exchange with the ambience. Heat exchange calculations must consider direct and reflected infrared radiation that may origin from natural and artificial infrared sources such as the sun, heating and heated devices, light bulbs and living bodies.

3.4 Common Challenges to Maintaining Core Stability

Internal and external conditions may challenge the maintenance of core temperature within a narrow range. Heat gain may arise from internal or external causes, while heat loss is mainly determined by external conditions.

3.4.1 Internal Challenge

Heat is a by-product of all metabolic reactions and temperature is the intensity measure of thermal energy. Typically, the heat generation from general metabolism and heat produced by skeletal muscles is weighted differently in thermoregulation.

3.4.1.1 Metabolism

Before mammalian cells become committed to growth and proliferation, it seems logical that two key criteria should be satisfied:

(1) that sufficient nutrients are available to provide the raw materials for biosynthesis and
(2) that sufficient energy is available to drive the process [56].

Metabolism is broadly defined as the sum of biochemical processes in living organisms that either produce or consume energy.

It is a dauntingly large sum: more than 8,700 reactions and 16,000 metabolites are now annotated in the *Kyoto Encyclopedia of Genes and Genomes* [57]. In thermal physiology **metabolism** invariably relates to the transformation of chemical energy into work and heat. In other divisions of physiology; the term is used in relation to other changes in state of chemical energy (e.g. from storage compounds to readily available compounds, including the energy that is transferred by polyphosphates, and vice versa), or to the turnover of material between different states (e.g. calcium salts from ionized to solid, and vice versa) [2].

Almost all the nutrient energy intake of an individual organism is used for various physiological needs; only a small proportion is lost in the faeces and urine. Obligatory energy expenditure is the energy used for special physiological needs such as growth in all organs, and in the reproductive organs during pregnancy or lactation. A variable proportion of the energy intake is used in physical activity; and most of ingested calories is used for metabolic processes or is lost in the production of heat. Obligatory thermogenesis refers to the minimal heat produced by all the processes that maintain the body in a basal state (fasting) at thermoneutral temperature. The standard metabolic rate and the heat generated during digestion, processing and storing of energy in the organs of the gastrointestinal tract including the liver and white fat, contribute to obligatory heat production [57].

Based on the function and tissues of heat production, thermogenesis can be further classified into seven categories: standard metabolic rate, thermal effect of food, cold-induced shivering thermogenesis, cold-induced non-shivering thermogenesis, diet-induced thermogenesis, non-exercise activity thermogenesis, and thermal and work effect of exercise.

Adaptive thermogenesis and movement-related heat and energy expenditure are subcategories of facultative thermogenesis, which have a major role in defence against cold. Adaptive thermogenesis is defined as regulated heat production in response to environmental temperature or diet. There are three subcategories of adaptive thermogenesis. Cold exposure induces shivering thermogenesis in skeletal muscle, and non-shivering thermogenesis in brown fat. Although current evidence does not indicate a role of muscle in non-shivering thermogenesis, indirect evidence suggests that such mechanisms may exist. Overfeeding triggers diet-induced thermogenesis; this is also a function of brown fat [57].

The "Glossary of Terms for Thermal Physiology" lists some definitions related to metabolism, but they are generally defined as they are targeted to the thermal physiology of animals and humans [2] (Table 3.12).

It is important to understand that all terms related to metabolism define conditions of energy production or expenditure. They do not provide direct temperature information, even in the case of thermal energy. The relationship between temperature and energy can be found in the definition of the unit Joule: 1 Joule is expressed as the 1/4.184 part of heat energy required to raise the temperature of a unit weight (1 g) of water from 0°C to 1°C. 4.184 Joules are equal to the traditional unit of 1 calorie.

Table 3.12 Terms related to metabolism

Metabolism, anaerobic

Transformation of matter and energy without uptake of oxygen [2].

Metabolic energy production. →Metabolic rate (*M* or *MR*): = metabolic energy transformation.

[W], [$W \cdot m^{-2}$], [$W \cdot m^{-3}$], [$W \cdot kg^{-1}$], [$W \cdot kg^{-3/4}$] The rate of transformation of chemical energy into heat and mechanical work by aerobic and anaerobic metabolic activities within an organism, usually expressed in terms of unit area of the total body surface, i.e. as a **heat flux**. The quantity ***M*** in the **body heat balance equation** [2].

Note: **Metabolic energy transformation** may not all result from aerobic metabolic activities and may therefore exceed that indicated by the rate of oxygen consumption. Part of the **metabolic energy transformation** may be used to do work on an external system, and therefore the rate of heat production may be less than the **metabolic energy transformation**. Terms in the **body heat balance equation** are often expressed as quantities of **energy** per unit surface area and per unit time [$W \cdot m^{-2}$], because heat exchange is a function of area. **Metabolic rate (*M* or *MR*)** is given as the total energy production in the organism in unit time [W] or often as the energy production per unit mass of tissue in unit time [$W\ kg^{-1}$]. For comparison of **metabolic rates** of animals of different body sizes, **metabolic rate (*M* or *MR*)** is usually related to (body mass)3/4 (→**metabolic body size**). [2]

Metabolic body size

The function of an animal's body size to which **standard metabolic rate (*SMR*)** (or **basal metabolic rate (*BMR*)**) is directly proportional. Synonym: **metabolically effective body mass**.

Surface rule

A statement that the **basal metabolic rate (*BMR*)** is proportional to the 2/3 power of body mass. [2].

Note: The rule is based on the proposition that **basal metabolic rate (*BMR*)** is related to surface area and that surface area varies with the 2/3 power of body mass. However, this is not experimentally verifiable, for when **basal metabolic rate (*BMR*)** is expressed per 2/3 power of body mass it increases systematically with body size (Kleiber, 1947). **Basal metabolic rate (*BMR*)** is more nearly proportional to the 3/4 power of body mass (→**metabolic body size**). Reference: Kleiber M, *Physiological Reviews* 1947 27, 411 [2].

Metabolically effective body mass: →metabolic body size (preferred synonym).

Note: The term **metabolically effective body mass** may be wrongly understood to be that part of the body mass which is metabolically active in contrast to a part that is metabolically inert. Its use should be avoided [2].

Metabolic rate, basal (*BMR*)

metabolic energy transformation calculated from measurements of heat production or oxygen consumption in an organism in a rested, awake, fasting* sufficiently long to be in post-absorptive state, and **thermoneutral zone** (a particular case of **standard metabolic rate (*SMR*)**). [W], [$W \cdot m^{-2}$], [$W \cdot m^{-3}$], [$W \cdot kg^{-1}$], [$W \cdot kg^{-3/4}$] [2].

Note: In these conditions, when the amount of work being done on an external system is negligible, the rate of heat production is equal to the rate of **metabolism (metabolic energy production)** [2].

* The period of fasting needs to be specified as this may be for days in large animals, and for much shorter periods for very small mammals and birds [2].

Metabolic rate, lowest observed (*LOMR*)

The lowest observed rate of **metabolism** during specified periods of minimum activity. [W], [$W \cdot m^{-2}$], [$W \cdot kg^{-1}$], [$W \cdot kg^{-3/4}$] →**metabolic rate, minimum observed (*MOMR*)**. The rationale and objective of **lowest**

(*Continued*)

Table 3.12 *(Continued)*

observed metabolic rate (*LOMR*) and **minimum observed metabolic rate (*MOMR*)** are identical: to measure the **metabolic rates** of small and wild animals during periods of minimal activity as the nearest that can be made to a measurement of a **standard metabolic rate (*SMR*)** [2].

Note: There are small but significant differences in technique: **Minimum observed metabolic rate (*MOMR*)** is the average **metabolic rate (*M* or *MR*)** during periods of minimum activity; **lowest observed metabolic rate (*LOMR*)** is the *lowest* recorded **metabolic rate (*M* or *MR*)** during periods of minimum activity. Such brief low values may be influenced by physical characteristics of the system of measurement. *Note:* **lowest observed metabolic rate (*LOMR*)** may be lower than **basal metabolic rate (*BMR*)** (e.g. during some stages of sleep) [2].

Metabolic rate, maximum (*MMR*)

The highest **metabolic rate (*M* or *MR*)** during a specified period of work compatible with sustained aerobic metabolism (i.e. when there is no progressive accumulation of lactic acid in the blood). [W], [$W \cdot m^{-2}$], [$W \cdot m^{-3}$], [W·kg–1], [$W \cdot kg^{-3/4}$] [2].

Note: There may be some confusion between **maximum** and **peak metabolic rates (*PMR*)**. Both terms indicate maximum rates. The distinction is in usage: **maximum** relates to work, **peak** relates to cold exposure. The terms should also be distinguished from **maximum oxygen consumption** [2].

Metabolic rate, minimum observed (*MOMR*)

Averaged **metabolic rate** during specified periods of minimum activity. [W], [$W \cdot m^{-2}$], [$W \cdot m^{-3}$], [$W \cdot kg^{-1}$], [$W \cdot kg^{-3/4}$] [2].

Note: The **metabolic rate (*M or MR*)** of small animals in particular, but also of larger wild animals, cannot be measured under basal or other standard conditions (→**metabolic rate, basal (*BMR*)**, and **metabolic rate, standard (*SMR*)**). A practical solution to the problem is to measure **metabolic rate (*M* or *MR*)** continuously and accept the average **metabolic rate (*M* or *MR*)** during periods of minimum activity as the best possible estimate of **standard metabolic rate (*SMR*)** [2].

Metabolic rate, peak (*PMR*)

The highest **metabolic rate (*M* or *MR*)** that can be induced in a resting animal by any cold environment. [W], [$W \cdot m^{-2}$], [$W \cdot m^{-3}$], [$W \cdot kg^{-1}$], [$W \cdot kg^{-3/4}$] Synonym: **summit metabolic rate→metabolic rate, maximum (*MMR*)** [2].

Note: Although summit metabolic rate (*SMR*) is now an accepted term, **peak metabolic rate (*PMR*)** is preferable because the abbreviation *SMR* is indistinguishable from that for **standard metabolic rate (*SMR*)** [2].

Metabolic rate, resting (*RMR*)

The **metabolic rate (*M* or *MR*)** of an animal that is resting in a thermoneutral environment but not in the post absorptive state. [W], [W·m^{-2}], [W·m^{-3}], [W·kg^{-1}], [W·kg$^{-3/4}$]. [2].

Note: A particular case of **standard metabolic rate (*SMR*)** used when the subject cannot be brought to a fasting condition, e.g. ruminant animals. The period of food deprivation should be stated. [2].

Metabolic rate, standard (*SMR*)

Metabolic energy transformation calculated from measurements of heat production or oxygen consumption in an organism under specified standard conditions*. [W], [W·m^{-2}], [W·m^{-3}], [W·kg^{-1}], [W·kg$^{-3/4}$] [2].

Note: The conditions are usually such that the amount of work being done on an external system is negligible. The rate of heat production is then an acceptable index of the rate of metabolism. [2].

* The specified standard conditions are usually that the organism is rested (or as near to rested as is possible), fasting (if possible), awake, and in a thermo neutral environment. The extent to which standard conditions can be achieved varies with species. →**Metabolic rate, minimum observed (*MOMR*)**[2].

Metabolic level: The **metabolic heat production** measured under standard conditions (→**metabolic rate, standard (*SMR*)**) during a 24 h period divided by the **metabolic body size**.
[kJ·kg$^{-3/4}$·(24 h)$^{-1}$]* [2].

Note: *1 kJ.kg$^{-3/4}$ · (24 h)$^{-1}$ = 0.2388 kcal · kg$^{-3/4}$·(24 h)$^{-1}$. **Metabolic levels** are approximately constant within phylogenetic groups but may vary between groups. For example, mammalian and avian species (**tachymetabolism**) have higher **metabolic levels** than other species (**bradymetabolism**), while the **metabolic levels** of birds are apparently higher than those of mammals, and those of prototherian (monotreme) and metatherian (marsupial) mammals are lower than those of eutherian (placental) mammals. (Poczopko, 1971). Reference: Poczopko P, *Acta Theriologica* 1971, 16, 1 [2].

Metabolic scope. →**Scope, metabolic**

Metabolic rate (*M or MR*) at **maximum oxygen consumption** divided by **least observed metabolic rate (*LOMR*)**. [2].

(*Continued*)

Table 3.12 (*Continued*)

Metabolic heat production ($\dot{H}$)

The rate of transformation of chemical energy into heat in an organism, usually expressed in terms of unit area of the total body surface (→**total body area**). $\dot{H}$ is equal to M – (W) in the **body heat balance equation**. [$W \cdot m^{-2}$], [$W \cdot kg^{-1}$] or [W]

Note: During **positive work rate** (+W) or in the absence of work (W = 0) **metabolic heat production** ($\dot{H} = M - (+W)$), i.e. **positive work rate** subtracted from **the rate of metabolic energy transformation (=metabolic rate)**, equals **total heat production**. When work is being done on the body by an external source (**negative work**), **total heat production** is the sum of **metabolic rate (*M* or *MR*)** (which is equal to metabolic heat production in this condition) and the heat liberated within the body due to **negative work** (M– (– W)) [2].

Total heat production

The rate of transformation of chemical energy into heat in an organism (**metabolic heat production ($\dot{H}$)**) plus any heat flow liberated within the body resulting from work done on the organism by an external force (**negative work rate**). [W], [$W \cdot m^{-2}$], [$W \cdot m^{-3}$], [$W \cdot kg^{-1}$] [2].

Note: During **positive work** and when no work is being done on or by the organism, **total heat production** equals **metabolic heat production ($\dot{H}$)** [2].

Metabolism, energy: The sum of the chemical changes in living matter in which energy is transformed. (Greek: *metabole*-change) [2].

Area, total body (A_b)

The area of the outer surface of a body, assumed smooth. [m^2] [2].

Note: Direct measurements of surface area are difficult and subject to appreciable error. Surface area is usually estimated from a formula such as that of Meeh (1879) that relates **total body area (A_b)** to body weight (W): $A_b = k \cdot W^{2/3}$. This is a particular case of Huxley's allometric law. Estimates of the value of k vary widely between and within species but are generally between 0.07 and 0.11 when A_b is in m^2 and W is in kg (Spector, 1956). The DuBois formula (→**Area, DuBois**), used in the estimation of the **total body area**, relates **total body area** to both mass and height. Estimates of **total body area** made with either Meeh's formula or the DuBois formula have limited accuracy, and reference should be made to the original direct measurements from which **total body area** was calculated. References: Huxley JS, *Problems of Relative Growth*, London: Methuen, 1932; Meeh K, Z. *Biol.* 1879 15, 425; Spector WS, *Handbook of Biological Data*, Philadelphia, PA: Saunders, 1956 [2].

Area, DuBois (A_D)

The total surface area in square meters of a nude human as estimated by the formula of DuBois based on the height, H [m], and mass, W [kg]. Reference: DuBois EF, *Basal Metabolism in Health and Disease*, Philadelphia, PA, Lea and Febiger, 1927.
$A_D = 0.202 \cdot W^{0.425} \cdot H^{0.725}$ [m²] [2].

Work rate, negative (−*W*)

The **rate** of work **(→power)** done on an organism by an external force. The quantity (−*W*) in the **body heat balance equation**. [[W], $W \cdot m^{-2}$], [$W \cdot m^{-3}$], or [$W \cdot kg^{-1}$] Antonym: **work rate, positive** [2].

Work rate, positive (+*W*)

The rate of work **(→power)** done by an organism on an external system. The quantity (+*W*) in the **body heat balance equation**. [W], [$W \cdot m^{-2}$], [$W \cdot m^{-3}$], or [$W \cdot kg^{-1}$] Synonym: mechanical energy, **work production, useful work accomplished**. Antonym: **work rate, negative** [2].

Power (W)

Physically power [W] is **energy** per time unit. In the case of mechanical energy per time unit it is called mechanical power or, particularly in thermal, sports and occupational physiology, **work rate** and attributed the letter *W*. [2].

Note: W, e.g. in the **heat balance equation**, is work rate in [W] and not work in [Ws = J]. In the case of thermal energy (heat) per time unit it is called heat flow and attributed the letter *H* [W]. [2].

Note: All quantities in the so-called **body heat balance equation** constitute power (either heat **flow** or **work rate**). For convenience, they may be expressed in [W], in [$W \cdot m^{-2}$], in [$W \cdot m^{-3}$], or in [$W \cdot kg^{-1}$] [2].

Work efficiency (η)

Work done on an external system per unit of energy expended by an organism in the performance of that work (i.e. total energy expended by an organism during the performance of work less that of **basal metabolism**). [%] [2].

3.4.1.1.1 *Basal metabolic rate (BMR)*

BMR is an index of the general level of activity of an individual's body metabolism, determined by measuring his oxygen intake in the basal state.

All BMR measurements must meet the following conditions [58]:

1. The subject should be completely rested, both before and during the measurements. They should be lying down and fully awake.
2. The subjects should be fasted for at least 10–12 h before the measurements are taken.
3. The environment in which the measurements are taken should be thermoneutral (22–26°C) so that there is no thermoregulatory effect on heat production.
4. The subject should be free from emotional stress and familiar with the apparatus used.

Depending on age and lifestyle, BMR represents 45% to 70% of daily total energy expenditure, and it is determined mainly by the individual's age, gender, body size and body composition.

The primary purpose of collecting BMR measures during the early part of the 20th century was to diagnose hypo- and hyperthyroidism. In recent years, BMR measures have been used to estimate total energy requirements or to better understand energy regulation.

Energy requirement is the amount of food energy needed to balance energy expenditure to maintain body size, body composition and a level of necessary and desirable physical activity consistent with long-term good health. This includes the energy needed for the optimal growth and development of children, for the deposition of tissues during pregnancy, and for the secretion of milk during lactation consistent with the good health of mother and child.

The recommended level of dietary energy intake for a population group is the mean energy requirement of the healthy, well-nourished individuals who constitute that group.

From the beginnings of BMR measurements, equations were developed to predict the basal metabolic rate based on surface area, body weight and body height (Table 3.13). However, the equations do not predict BMR accurately and overestimate more often than underestimate the basal metabolic rate.

Table 3.13 Equations for predicting basal metabolic rate [58]

Author (year of publication)	Equation
DuBois and DuBois (1915)	$A = W^{0.425} \times H^{0.715} \times 71.84$ A = surface area in cm^2; W = weight in kg; H = height in cm
Harris and Benedict (1919)	Men: $h = 66.4730 + 13.7516W + 5{:}0033S - 6.7750A$ Women: $h = 665.0955 + 9.5634W + 1.8496S - 4.6756A$ h = kcal day^{-1}; W = weight in kg; S = stature in cm; A = age in years
Quenouille et al. (1951)	Men: $M = (2.975H + 8.90W + 11.7S + 3.0h - 4.0t) + 293.8$ M = kcal day^{-1}; W = weight in kg; H = height in cm; S = surface area from DuBois; t = temperature, h = humidity
Schofield equations (WHO/FAO/UNU) (1985)	Men 10–18 years BMR = $69.4W + 322.2H + 2392$ 18–30 years BMR = $64.4W - 113.0H + 3000$ 30–60 years BMR = $47.2W + 66.9H + 3769$ > 60 years BMR = $36.8W + 4719.5H - 4481$ Women 10–18 years BMR = $30.9W + 2016.6H + 907$ 18–30 years BMR = $55.6W + 1397.4H + 146$ 30–60 years BMR = $36.4W - 104.6H + 3619$ > 60 years BMR = $38.5W + 2665.2H - 1264$ W = weight in kg; H = height in m; BMR = kcal day^{-1}
Mifflin et al. (1990)	Men: BMR = $9.99W + 6.2\,H - 4.92$age + 5 Women: BMR = $9.99W + 6.25H - 4.92$age − 161 W = weight in kg; H= height in m; BMR = kcal day^{-1}
Oxford equations (2005)	Men 10–18 years BMR = $15.6W + 266H + 299$ 18–30 years BMR = $14.4W + 313H + 113$ 30–60 years BMR = $11.4W + 541H - 137$ > 60 years BMR = $11.4W + 541H - 256$ Women 10–18 years BMR = $9.4W + 249H + 462$ 18–30 years BMR = $10.4W + 615H - 282$ 30–60 years BMR = $8.18W + 502H - 11.6$ > 60 years BMR = $8.52W + 421H + 10.7$ W = weight in kg; H = height in m; BMR = kcal day^{-1}

3.4.1.1.1.1 *The metabolic equivalent (MET)*

The metabolic equivalent (MET) is a widely used physiological concept that represents a simple procedure for expressing energy cost of physical activities as multiples of resting metabolic rate (RMR). The MET is an assigned unit of measurement to designate "sitting-resting" metabolic rate of man.

$$1 \text{ MET} = 58.15 \text{ W·m}^{-2} = 50 \text{ kcal·h}^{-1}\text{·m}^{-2}$$

It is an empirical unit of measurement to express the **metabolic rate (*M* or *MR*)** of a man whose clothing has an insulation value of 1 clo when he is sitting at rest in comfortable indoor surroundings (21°C) [2].

Alternatively, 1 MET is expressed as 3.5 ml O_2 $kg^{-1}·min^{-1}$ or 1 kcal kg^{-1} h^{-1}, and this value was first derived from the resting O_2 consumption ($\dot{V}O_2$) of one person, a 70 kg, 40-year-old man [59]. A study investigated the resting metabolic activity in a cohort of 642 women and 127 men, 18–74 years of age, 35–186 kg in weight, who were weight stable and healthy, albeit obese in some cases. The results showed that average VO_2 and energy cost corresponding with rest (2.6±0.4 ml O_2 kg^{-1} min^{-1} and 0.84±0.16 kcal $kg^{-1}h^{-1}$, respectively) were significantly lower than the commonly accepted 1-MET values of 3.5 ml O_2 kg^{-1} min^{-1} and 1 kcal kg^{-1} h^{-1}, respectively. Body composition (fat mass and fat-free mass) accounted for 62% of the variance in resting VO_2 compared with age, which accounted for only 14% [60].

3.4.1.1.2 *Thermic effect of food*

Oral or parenteral administration of nutrients in healthy individuals is associated with a stimulation of energy expenditure. This increase over baseline of energy expenditure associated with feeding is referred to as the thermal effect of food (TEF). In healthy subjects over a 24 h period, TEF represents about 10% of the total amount of energy expended. Calculation of the thermal effect of food is based on calorimetric monitoring of energy dissipation during administration, absorption and disposal of a bolus of nutrients until energy expenditure returns to baseline values [61]. TEF may also be labelled as diet-induced thermogenesis as it was defined as the increase in energy expenditure above basal fasting level divided by the energy content of the food ingested and is commonly expressed as a percentage [62]. The main determinant of TEF is the energy content of the food, followed by the protein fraction of the food. The thermal effect of alcohol is like the thermal effect of protein [62].

3.4.1.1.3 *Diet-induced thermogenesis*

Diet-induced thermogenesis was described over a century ago in Germany as "luxury-consumption"; that is, a physiological mechanism exists that permits excessive calorific intake to be dissipated as heat, allowing individuals to eat without gaining weight [57]. Overfeeding triggers diet-induced thermogenesis in animals, where the involvement of brown fat is well supported. In humans, the contribution of brown fat to diet-induced thermogenesis in humans is yet not totally understood.

3.4.1.1.4 *Cold-induced non-shivering thermogenesis*

It is now established, that brown fat plays an important role in non-shivering thermogenesis not just in small animals, but also in humans beyond the perinatal period.

3.4.1.1.4.1 *Brown adipose tissue*

Brown fat is a form of adipose tissue with selected distribution. It differs from white adipose tissue both structurally and functionally; its colour is brownish (yellow to reddish), depending on its content of (mitochondrial) cytochromes and of fat that is distributed in multiple droplets (multi-locular) as compared to white fat cells (mono-locular) and on its dense vascularization. Brown fat appears to be restricted to mammalian species, especially of smaller body size, including the hibernators, but is also present in newborns of larger species including humans. Regulatory heat production has been identified as its main but possibly not its only function. Its heat-generating capacity relates to the rate of fatty acid oxidation that is not constrained by cellular ATP catabolism. This property correlates with the presence of a specialized component in the inner mitochondrial membrane named **UCP1** (the **uncoupling protein** or **thermogenin**, which establishes a proton shunt bypassing oxidative phosphorylation). The activity of brown fat is controlled by the sympathetic nervous system through release of adrenaline [2].

Chronic exposure to cold and to excess food intake enhances the heat-generating capacity of **brown fat** by hypertrophy (including increase in **UCP1** amount) and hyperplasia. The total process is referred to as recruitment. In accordance with the sympathetic control of **brown adipose tissue**, the increase in **metabolic energy transformation** of an animal in a **thermoneutral** environment after systemic injection of noradrenaline is considered as an estimate of the thermogenetic capacity of **brown adipose tissue [2]**.

Cannon and Nedergaard reported that in small animals the heat-producing capacity of brown fat can be calculated to be some 300 W/kg when it is working at its highest intensity [50]. This is the 3.5-fold capacity of thyroid tissue and almost 600 times of the heat generation rate of white fat cells [49]. However, the few in vivo measurements of brown fat temperature reported depended on the stimulus temperature, increases between 0.2 and 1.0°C [63–66]. Based on micro-calorimetry, an individual brown fat cell produces 1 nano joule of thermal energy during stimulation by 1 μm noradrenaline and thus leading to an increase of 0.27°C in surface temperature of the adipose cell [67]. As temperature is an intensive quantity meaning the level of temperature is independent of mass, stimulation of 100 or 100,000 cells will result in the same temperature increase at the surface.

3.4.1.2 Heat production by skeletal muscles

Skeletal muscles play an important role in biological heat generation. Archibald Vivian Hill received the Nobel Prize in 1923 for his discovery relating to the production of heat in the muscle with very subtle temperature changes associated with muscle contraction.

Hill observed that the rate of heat produced by a shortening muscle is proportional to the shortening velocity (rate of heat production = aV). He showed experimentally that the rate of heat production (aV) plus the rate of mechanical work performed ($F \bullet V$) depended linearly on the load.

$$(aV + FV) = (F_0 - F)b,$$

where F is force, F_0 maximum isometric force, V shortening velocity, and a and b constants.

During isotonic contraction is more liberated than during isometric contraction. Fenn called this extra heat the shortening heat.

In this case, the shortening heat is proportional to the shortening of the muscle; the larger the shortening distance, the more extra heat is produced (this is called the Fenn effect) [68].

Meyerhof and Lohmann have shown that creatine phosphoric acid has a large heat of hydrolysis, about 11,000 calories per gram molecule, but the total heat of contraction of a muscle has the same value relative to the lactic-acid production. In 1934, Lohmann showed that muscle extracts adenosine-triphosphate (ATP) and phosphocreatine (PCR) were held in equilibrium by a specific enzyme now known as creatine kinase.

In the muscle cell, thermogenesis occurs through a series of uncoupled reactions. Neurotransmitter-mediated opening of cell-surface Na+ channels leads to release of Ca^{2+} into the cytoplasm from sources both outside the cell and the sarcoplasmic reticulum via the ryanodine receptor. Malfunction of this receptor leads to uncontrolled Ca^{2+} release, and that explains the thermogenesis in malignant hyperthermia.

Ca^{2+} results in heat generation from ATP hydrolysis during both muscle relaxation and actin–myosin cross-bridge cycling during sustained contraction. Additional heat energy is released when Ca^{2+} ions are pumped back into the sarcoplasmic reticulum [57].

In resting states, intramuscular temperatures are not homogeneous, but increase with sensor depth and vary with proximity to large arteries and bones [69].

Figure 3.10 Provides examples of in vivo resting muscle temperatures that are in general below the core temperature of the human body. Among muscles, however, the inter-site variability expands as metabolic variations from one segment to another increase [69].

3.4.1.2.1 *Cold-induced shivering thermogenesis*

Shivering is a type of muscular contraction in which mechanical efficiency is very low. For this reason, the energy generated by shivering appears almost entirely in the form of thermal energy. The role of shivering in thermoregulation was recognized a century ago when it was thought that shivering is a process of thermal regulation that produces heat by generalized contraction of muscles. Its origin is reflex, but it may be partly subjective and temporarily under the control of the will.

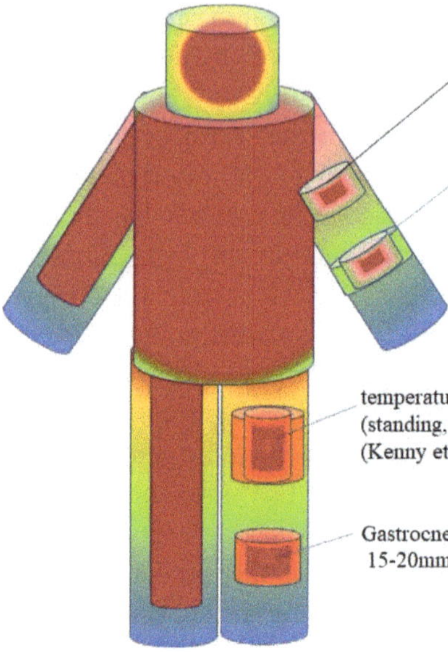

Figure 3.10 In vivo muscle temperatures.

Muscular contractions during shivering are of short duration but are repeated, appearing as successive bursts. The intensity of shivering is related to the number of contracting muscles (spatial summation) and to the frequency of the contraction bursts (temporal summation). These contractions occur without movement of the limbs; which explains the low level of mechanical efficiency. Shivering is an important means of thermogenesis, at least in man, direct measurement of the heat produced is a good means of evaluation. More details on shivering thermogenesis can be found in Section 3.5.

3.4.1.2.2 *Non-exercise activity thermogenesis*

Skeletal muscles represent the engine that promotes mobility. The International Classification of Functioning, Disability and Health (ICF) defines mobility as one of nine domains that describe the health category activities and participation [70]. The subdomains include

1. changing and maintaining body position;
2. carrying, moving and handling objects;

3. walking and moving;
4. moving around using transportation.

Such forms of physical activity are not understood as exercise, although they need energy to fuel the muscular function and thus generating heat. Non-exercise activity thermogenesis is highly variable and ranges from ~15% of total daily energy expenditure in highly sedentary individuals to more than 50% in highly active people. Non-exercise activity thermogenesis is also generally higher in lean than obese individuals. Indeed, lean and obese individuals are different in the energy dedicated to non-exercise activity thermogenesis—on average 350 kcal per day [71].

3.4.1.2.3 *Thermic and work effect of exercise*

The experimental quantification of heat production in human skeletal during exercise is difficult, time-consuming and expensive as it is not sufficient to just measure temperature inside the muscles and estimate energy expenditure. The complexity of such research was elegantly demonstrated in a paper from Denmark [72].

At the start of exercise, rapid intramuscular temperature changes are observed), reflecting variations in heat storage, with temperatures becoming more uniform as exercise continues. Gonzalez-Alonso et al. quantified the total skeletal muscle heat production in humans performing intense work during dynamic knee extension exercise by summing the heat storage in the contracting muscles, the heat exchange with the body core by the circulation, and heat release to the skin and environment [72]. An increase of muscle temperature of approximately 1.1°C has been observed at the end of the exercise. Muscle is not only a significant heat source during exercise, but functions also as an important heat sink.

3.4.2 External Challenge

The ambient air temperature on earth has a wide range from -60°C to +50°C and both extremes are a major challenge for maintaining core temperature of the living human body. Exposure of the human body to extremes of temperature results in harmful effects either localized or generalized dependent on the energy level. Human responses occur on voluntary and involuntary basis.

3.4.2.1 External heat load

The main external heat source is the infrared radiation from the sun, which was original phenomenon that led William Herschel to the detection of infrared rays. Man-made heat sources such as burning substances to create fire have been recognized from early civilization. Modern technology has developed multiple industrial and commercial processes that generate heat into man's surroundings.

3.4.2.2 External heat loss

In his natural states, man must be regarded as a living being that can survive only in a warm environment. Since the capacity of defence mechanisms against heat loss is very limited, survival became possible by developing behavioural defence mechanisms such as seeking and building shelters, clothing and other means of protection.

3.5 Response to Challenge

The response to challenge is dependent on the temperature level, the anatomical site and the size of the challenged area and the duration of the stimulus. Details of thermoreceptors throughout the body are given in Section 3.2.1.1. It is important to note that the distribution of thermoreceptors in the skin is uneven. The receptive field of a peripheral thermal sensor has a diameter of 1 mm only. So, the spatial temperature resolution of natural thermoreceptors is much lower than the spatial resolution of a modern infrared detector array.

3.5.1 Defence against Heat Load

There are behavioural and autonomic responses to protect the body against overheating. They include the increase of the area of heat dissipation, reduction of blood perfusion due to vasodilation and activation of sweat glands to promote cooling by evaporation.

3.5.1.1 Increase of the area of heat dissipation

The principles of heat transfer by conduction, convection and radiation apply as discussed in the Chapter 2.

3.5.1.1.1 *Behavioural*

Examples of changes in posture to modify heat exchange to and from the human body are illustrated in Table 3.9. Removing clothing with high insulation or seeking shelter in the shade are self-explanatory.

3.5.1.1.2 *Autonomic*

Under resting conditions in a thermo neutral environment, core temperature is regulated by modifications of the cardiovascular system that is controlled by the autonomic nerve system. All temperature related responses are governed by the sympathetic nerves, which control the diameter of the blood vessels in the periphery. Acetylcholine is responsible for widening the blood vessels, while noradrenaline acts in reverse.

3.5.1.1.2.1 *Vasodilation*

The vessels of the cardiovascular system are the means of supplying blood according to demand of tissues and organs. Changes in blood distribution occur by the variation in the diameter of the vascular beds, while the volume of blood circulating in the cardiovascular system is roughly constant. Vasodilation refers to the widening of blood vessels. The process is the opposite of vasoconstriction, which is the narrowing of blood vessels. Both result from changes in smooth muscle cells within the vessel walls.

Blood flow is defined by the ratio of blood volume to tissue volume (ml 100 ml^{-1}) or tissue weight (ml 100 g^{-1}) respectively, while perfusion is the rate of change in blood flow expressed in ml min^{-1}100 ml^{-1} or ml min^{-1} 100 g^{-1}.

Variation of blood flow depends on certain factors such as the vasodilation of vessels, change in the distribution of the cardiac output and variation in the cardiac output itself. Vascular conductance is a term defined as perfusion divided by the mean arterial blood pressure (ml min^{-1} $(100\ ml)^{-1}$ $mmHg^{-1}$). Cutaneous vascular conductance reaches it maximal value, when the resistance vessels are totally relaxed [73].

Variation of the volume of blood being pumped by the heart per unit time is referred to as cardiac output. This is the product of the heart rate, the number of heart beats per minute, and the stroke volume, defined as the volume of blood pumped from the ventricle per beat.

The first step in heat dissipation is vasodilation of skin vessels. All cutaneous vessels are in the dermis and originate in the vascular hypodermal network, which is at a depth of about 2 mm. The physiological mechanisms of vasodilation triggered by an increase of core temperature or vasodilation caused by local skin heating are not the same.

3.5.1.1.2.1.1 Vasodilation due to increased core temperature

For a long time, it was believed that the mechanism of core temperature-induced vasodilation was mainly nervous. However, the mechanisms responsible for cutaneous active vasodilation in humans is now regarded as a much more complex picture than mediation by a single vasodilator transmitter such as acetylcholine. It appears to be due to multiple contributors, some acting through nitrogen oxide. The early portion of active vasodilation is apparently via acetylcholine, whereas the latter parts are not, but an important part is accomplished via co-transmitters released from sympathetic cholinergic nerves [73].

According to the current theory, active cutaneous vasodilation is based on the action of transmitter substances, released from sympathetic nerve terminals, on endothelial cells and smooth muscle cells of blood vessels [73]. Three transmitter substances acetylcholine (Ach), the vasoactive intestinal peptide (VIP) and pituitary adenylate cyclase-activating peptide (PACAP) are involved. Ach binds to a muscarinic receptor- 3 (M_3) at the endothelial cells, while PACAP connects to the pituitary adenylate cyclase-activating peptide 1 receptor (PAC_1) or the vasoactive intestinal peptide receptor 1 or 2 ($VPAC_1$ or $VPAC_2$ of smooth muscle cells. VIP binds also to the muscle cell via $VPAC_1$ or $VPAC_2$ and seems to act on mast cells to release histamine that binds via histamine-1 receptors (H_1) to endothelial cells.

Endothelial cells are the main source of nitric oxide (NO). M_3 and H_1 receptors and an eventual activation of the transient receptor potential vanilloid receptor 1 (TRPV-1) stimulate the intracellular neural NO synthase (nNOS) isoform leading to nitric oxide production. Prostaglandins may also have a role in active cutaneous vasodilation, a yet unknown transmitter is thought to activate the cyclooxygenase enzymes resulting in generation of prostacyclin (PGI_2).

The vascular smooth muscle cell is the target of all the above transmitters. The mechanism of vasodilation due to activation of $VPAC_1$, $VPAC_2$ or PAC_1 have not yet been determined, NO contributes up to 40% to 50% of active vasodilation. Nitric oxide activates a chain of second messengers including soluble guanylate cyclase (sGC) and cyclic guanosine monophosphate (cGMP), leading to a decrease in intracellular Ca^{2+} concentration and that results in the uncoupling of contractile proteins and vascular relaxation.

3.5.1.1.2.1.2 Vasodilation due to increased skin temperature

The vasodilatory response to local heating follows a different pathway than the response to an increase of core temperature. In 1943, Barcroft and Edholm showed that a non-linear relationship exists between blood flow in the forearm and local skin temperature. Larger changes in blood flow relative to skin temperature were observed above 37°C, reaching a maximum at 42C [74, 75]. Different vasomotor responses occur in painful or non-painful heating [76].

In addition to classical reflex arcs, the axon reflex is involved in vasodilation. Stimulation of nociceptive nerve fibres results in a triple response of the skin, characterized by a local capillary dilation, a wheal that follows it, and a surrounding flare. The secretion of peptides causing these phenomena is the result of the response within the nociceptive fibres to the sensed noxious stimulus.

The typical response of skin blood flow to a rapid local heating to 42°C was described as an immediate and steep increase to ~ 70% of the maximal cutaneous vascular conductance (CVC), a subsequent drop to 60% CVC and a further increase of vasodilation to a sustained plateau. The last phase is characterized by an attenuation of the plateau, the so-called "die-away" phenomenon [73].

The biochemical and physiological processes leading to vasodilation in the first and in the plateau-phase differ in the involved compounds and their interaction. In the first phase in which the axon reflex play an important role, sensory/nociceptive nerves are releasing substance P (SP) and calcitonin gene related peptide (CGRP) which bind via neurokinin-1 receptors (NK_1) and via CGRP-receptors to both endothelial and smooth muscle cells. Activation of endothelial NK1 by SP and CGRP leads at one side in production of endothelial derived hyperpolarization factors (EDHFs), and at the other side in the stimulation of NO production. The latter is also stimulated by activation of the transient receptor peptide TRPV1, the adenosine receptors A_1 and A_2, and β_2 receptors activated by norepinephrine (NE) and neuropeptide Y (NPY) released from sympathetic nerve fibres.

All nitric oxide dependent mechanisms result in an up-regulation of Ca^{++}. EDHFs act through stimulation of calcium-activated potassium (K_{Ca}) channels leading to hyperpolarization of the endothelial and the smooth muscle cells. Hyperpolarization of the endothelial cells spreads through gap junctions to the smooth muscle, and the increased membrane potential of either origin causes muscle relaxation and thus vasodilation.

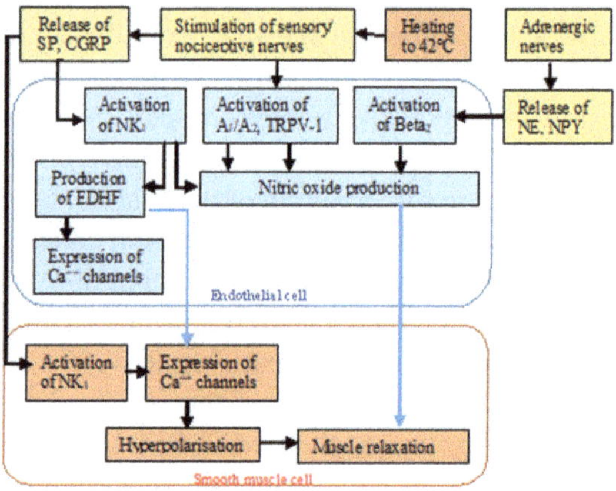

Figure 3.11 Mechanisms leading to the first increase of blood flow during rapid local heating to 42°C. Modified from [73], with permission from John Wiley and Sons.

The cellular mechanisms are slightly different during the plateau of cutaneous vascular conductance. While calcium-controlled channels and NO-related mechanisms contribute to equal parts to the peak phase, the inhibition of K_{Ca} results in a 25% reduction of vasodilation in the plateau phase, a further reduction to approximately 15% of cutaneous vascular conductance was observed after combined inhibition of both mechanisms.

3.5.1.1.3 *Cooling by water evaporation*

When the external heat gain is higher than the heat loss, the most rapid response is a rise in skin temperature. This rise allows an increased loss of heat by conduction, convection and radiation. If this response is not enough to achieve thermal balance, an increase in heat loss can only be achieved by raising the insensible heat loss that is heat loss by evaporation. In neutral conditions, evaporative heat loss is small. It takes the form of insensible perspiration from the skin and the respiratory tract. The rate of this evaporative heat loss can be termed *basal* and be symbolized by E_b.

3.5.1.1.4 *Heat loss by sweating*

In humans, effective heat loss is achieved by the secretion and evaporation (E) of sweat. According to the laws of thermodynamics, the heat that must be lost to achieve a new thermal balance has to be equal to the value of the heat storage (S). Therefore, we must have $E = S$ or expressed in terms of heat transfer; total heat transfer by evaporation (H_e) is equal to the sum of metabolic heat rate (H_m), basal evaporation (E_b) and evaporation (E):

$$H_e = H_m + E_b + E$$

The latent heat of vaporization of sweat is close to that of water itself: 2.43 kJ.g^{-1} of water (0.585 kcal). However, the value of latent heat of sweat varies in the literature between 2,398 J/g to 2,595 J/g [97]. Heat is removed from the medium in which the vaporization occurs, and in this way the medium tends to be cooled. Therefore, the vaporization of sweat must occur within the peripheral layers of the skin, or on the skin itself, to cool the body. If evaporation cannot be completed within these structures, sweat flows over the skin, resulting in a corresponding absence of evaporative heat loss.

Cooling of the body by sweating implies two mechanisms:

1. The production and secretion of sweat. This mechanism is essentially biological.
2. The evaporation of the sweat produced. This is mainly dependent on the physical characteristics of the ambient air and is thus mainly physical.

The thermoregulatory response is the main, but not the only cause of sweat production. Many diseases are associated with sweating and may lead to generalized, regional or focal hyperhidrosis due to increased sweat gland activity [77] (Table 3.14). The thermoregulatory function is rarely compromised in pathological sweating.

Table 3.14 Causes of hyperhidrosis

Primary	Focal	Excessive, bilateral, and relatively symmetric sweating occurring in at least one of the following sites: the axillae, palms, soles, or craniofacial region.; genetic background in some cases
Secondary	Generalized	drugs, substance abuse, cardiovascular disorders, respiratory failure, infections, malignancies, endocrine/metabolic disorders (thyrotoxicosis, hypoglycaemia, pheochromocytoma, acromegaly, carcinoid tumour), neurologic (Parkinson's disease)
	Regional	stroke, peripheral nerve damage, central or peripheral nervous system lesions that cause localized anhidrosis can cause compensatory sweating in other areas (stroke, spinal cord lesion, neuropathy, Ross syndrome)
	Focal	Frey syndrome, gustatory sweating, eccrine nevus, social anxiety disorder, unilateral focal hyperhidrosis (e.g. neurological disorder or tumour)

Note: Modified from 77, 78.

Heavy sweating during exercise in the heat affects the balance of fluids and minerals of the body, and consequently all related physiological control mechanisms. These interactions and the possible involvement of sweat in the pathophysiology of non-sweat related skin disorders is beyond the scope of this chapter.

3.5.1.1.4.1 *The sweat glands*

Sweat is produced by the *sudoral glands.* Based on the mechanism of sweat production, three types of sweat glands can be differentiated: eccrine, apocrine, and apo-eccrine glands [79].

3.5.1.1.4.1.1 Eccrine glands

1.6 to 5 million of these true regulatory sweat glands are distributed over the entire skin surface, although unevenly. They first appear in the 3-month-old foetus, on the volar surface of the hands and feet, and later appear in other areas of the body with some exceptions such as the lips, external ear canal and parts of the external genitals [80]. The highest sweat gland density, i.e. 700 sudoral glands per cm^2, can be found on the palms and soles and the least, i.e. 65 sudoral glands per cm^2, on the back. However, data on sweat gland density from different studies vary a lot. This is caused by different methodologies in the identification of sweat glands and possible variation between individuals.

3.5.1.1.4.1.1.1 Microscopic structure and physiology

An eccrine sweat gland consists of a single tubule ranging from 4 to 8 mm in total length that shows three portions [79]. The intra-epidermal part with a luminal diameter between 20 and 60 μm is called *acrosyringium.* It is followed by the *dermal duct* with a straight and a coiled part. The inner diameter of the duct is only 10–20 μm, the outer diameter is about 50–80 μm and the length of the straight duct is approximately 2 mm. Situated in the deep dermis, the coiled *secretory portion* is the largest part of the gland with an overall coil diameter of 500–700 μm and an inner diameter of the secretory tubule of 30–40 μm [79].

3.5.1.1.4.1.1.1.1 *The dermal duct*

The dermal duct is described as an epithelium of two to three layers of epithelial cells. They constitute a functional barrier between the luminal and extra cellular compartments.

The basic function of the sweat duct is the reabsorption of ions from primary sweat. This process is driven by a Na/K-ATPase on the basolateral membrane, pumping sodium from the ductal cells into the interstitial fluid. The chloride channel cystic fibrosis transmembrane conductance regulator (CFTR) is abundantly expressed in the human re-absorptive duct and promotes almost the total electrical conductance for chloride. Mutations in the ubiquitous CFTR gene lead to cystic fibrosis (CF), a clinical condition in which ductal electrolyte reabsorption is highly affected, leading to increased sweat chloride levels. Therefore, sweat analysis and CFTR genetic testing play a central role in the diagnosis of cystic fibrosis [81].

3.5.1.1.4.1.1.1.2 *Secretory portion*

This coiled tubule is composed of three distinctive cell types: clear, dark, and myoepithelial cells. It was believed that the clear cells are responsible for the precursor sweat secretion produced from blood plasma and that the dark cells produce mucus [3]. The clear secretory cells contain abundant mitochondria and form basal infolding cell membranes and intercellular canaliculi with other neighbouring clear cells. Basal infolding cell membranes are rich in Na,K-ATPase activity which provides the driving force to the secretion of eccrine sweat [80].

Dark cells contain granules in their cytoplasm, which are positive with PAS (Periodic acid–Schiff) staining. The obstruction of the secretory lumen by PAS positive material might be the mechanical cause for the development of miliaria, a sweat retention syndrome.

Myoepithelial cells surround the secretory portion of eccrine and apocrine sweat glands. Acetylcholine and alpha-adrenergic agonists respectively induce contraction of myoepithelial cells in sweat glands and they provide mechanical strength to the hydrostatic pressure during sweat secretion. Myoepithelial cells and their progenitors may have a role in regeneration of sweat glands [82].

3.5.1.1.4.1.2 Apocrine sweat glands

Apocrine sweat glands are restricted to hairy body areas, as they open and secrete into the hairy canal. For this reason, these glands can only be found in the axilla, the mammary, perineal and genital region. The density of apocrine sweat glands is approximately a tenth of the density of eccrine sweat glands.

The tubule and secretory coil of apocrine glands are bigger than that of endocrine glands. The secretory coil consists of two different cell types: secretory and myoepithelial cells. The function of the latter cells remains unclear. Apocrine secretion takes place as apical budding-off from the luminal cells and is under adrenergic control. The presence of adrenergic receptors in human apocrine sweat glands has been questioned for long [79], but a group in Glasgow identified at the basolateral membrane of the secretory coil cells β_2-adrenergic receptors, β_3-adrenergic receptors, and muscarinic acetylcholine receptors localized to granules within the secretory cells [83]. However, the relationship between cholinergic and adrenergic stimulation of sweating is not yet totally understood.

The secretory function of these glands begins at puberty and they respond to emotional stimuli such as anxiety, pain or sexual arousal. Due to their location, little of the sweat produced by these glands is lost by evaporation and most remains on the skin surface. Consequently, their contribution to evaporative cooling is small.

A conceptual model of psychological sweating and the associated cutaneous response was developed by M. Harker [84]. Starting point is the response to the perception of acute psychological stress. The frontal cortex, the hippocampus and the amygdala are involved in processing such a stress response, which elicits the primarily cholinergic neural innervation of eccrine sweat glands and the adrenergic peripheral activation of apocrine sweat glands. Generalized eccrine and localized apocrine sweating results in perceivable sweating from the palmar and plantar skin surfaces, face and axilla, with concomitant odour generation from the axilla. Particularly in social situations, this cascade of sweat secretions and self-perceived awareness of sweat and malodour sets up a vicious cycle, in which increased anxiety can result in more sweating, which in turn leads to more anxiety.

Psychogenic stimuli activated predominately sweat glands at the glabrous skin while thermal stimuli act preferably at the non-glabrous surface. Both stimuli simultaneously applied, accounted for ~90% of the site-specific maximal number of activated sweat glands. During the normothermic psychological stimulation, sweating from the glabrous surface was elevated via a 185% increase in the number of activated glands within the first 60 s [85].

3.5.1.1.4.1.3 Apo-eccrine sweat glands

First described by Sato et al. in 1987, apo-eccrine sweat glands have been identified in the axilla and the anogenital region. Up to 50% of axillary sweat glands can be apo-eccrine and they are presumed to develop during puberty. Apo-eccrine glands consist of myoepithelial cells, eccrine secretory cells and apocrine secretory cells. The gland morphology is characterized by an irregularly shaped tubule [79].

In vitro, apo-eccrine glands show a greater responsiveness to cholinergic as well as adrenergic stimuli than eccrine glands. The overall sweat rate of apo-eccrine sweat glands is higher than that of other types of sweat glands suggesting that they contribute strongly to axillary sweating.

3.5.1.1.4.2 *Sweat gland innervation*

It is well established that the nerve structure of eccrine sweat glands is composed of non-myelinated class C fibres of the postganglionic sympathetic fibres originating from the paravertebral ganglia. Nerve cells from spinal cord segments T2–T8 supply the skin of the upper limbs; those from T1–T4, the face and eyelids; those from T4–T12, the trunk; and those from T10–L2, the lower limbs [86]. The fibres are distributed to the myoepithelial and secretory cells. However, although these fibres are undoubtedly sympathetic, the chemical mediator is acetylcholine. Sweating is blocked by atropine, but it is evoked by acetylcholine and parasympathetic like drugs.

Local administration of adrenaline can stimulate sweat secretion. It has been stated that cholinergic innervation is responsible for thermal sweat secretion, but adrenergic innervation affects" psychogenic secretion." However, the source of adrenaline seems to be rather humoral than neurogenic. Recent experiments support a cholinergic mechanism being responsible for both, thermal and psychogenic sweating [87].

3.5.1.1.4.3 *Sweat production*

In 1956, Bulmer and Forwell [88] suggested that sweat was produced by ultrafiltration of the blood plasma, resulting in a precursor or primary sweat fluid. The removal of some Na^+ ions and other substances by the duct cells resulted in the final secretory fluid. Nowadays, the molecular mechanisms by which secretory cells produce the precursor fluid and the re-absorption of ions in the dermal duct is much better understood than 60 years before [89]. This is also true for the hormonal and neural control of sweat production and its composition [90].

3.5.1.1.4.3.1 Secretion volume

Sweat secretion depends on the thermal requirements of the body, providing that the necessary water is available. Sweat secretion is almost zero (except for the small production of the apocrine glands) in cold and neutral conditions. Secretion begins and grows progressively greater as soon as the heat load increases.

A recent comprehensive overview provided a critical view on the methods of sweat collection and many quantitative data on trans-epidermal water loss and sweat secretion rates under various thermal conditions [91]. Daily insensible water losses for a standardized individual (surface area 1.8 m^2) will be 0.6–2.3 L, with the hands (80–160 $g \cdot h^{-1}$) and feet (50–150 $g \cdot h^{-1}$) losing the most, the head and neck losing intermediate amounts (40–75 $g \cdot h^{-1}$) and all remaining sites losing 15–60 $g \cdot h^{-1}$.

Humans have a large and widely variable capacity for the active secretion of sweat. Whole-body sweat losses in men can exceed 2 $L \cdot h^{-1}$ during competitive sport, with rates of 3–4 $L \cdot h^{-1}$ observed during short-duration, high-intensity exercise in the heat [91].

3.5.1.1.4.3.2 Chemical composition of sweat

Sweat from eccrine and apocrine glands differ in composition. Water accounts to 99% of eccrine sweat. Further components are sodium, chloride, potassium, calcium, magnesium, lactate, ammonia, amino acids, urea and bicarbonate. In addition, several proteins and peptides have been identified including the antimicrobial peptide dermcidin [79].

Apocrine sweat glands produce an oily, odourless fluid, containing proteins, lipids and steroids. Some apocrine proteins function as carrier proteins for volatile odour molecules. Apocrine secretions may be mixed with sebum, as both, apocrine and sebaceous glands open into the hair follicle [79].

Apoeccrine glands secrete an eccrine-like watery fluid. Although similar in ion and protein contents as eccrine sweat, the detailed composition of apoeccrine sweat remains unknown [79].

Since sweat glands selectively reabsorb electrolytes, the sodium and chloride composition of discharged sweat varies with secretion rate. Across whole-body sweat rates from 0.72 to 3.65 $mg \cdot cm^{-2} \cdot min^{-1}$, sodium losses of 26.5–49.7 mmol·L^{-1} could be expected, with the corresponding chloride loss being 26.8–36.7 mmol·L^{-1} [91].

3.5.1.1.4.3.3 Patterns of sweat production

For any given gland, sweat production appears to be intermittent [3]. Houdas and Ring described the following pattern of sweat gland recruitment at the onset of sweating: the first area to sweat is generally the forehead, followed in order by the upper arms, hands, thighs, feet, and, finally, back and abdomen. However, Taylor and Machado-Moreira concluded in their carefully performed analysis of the literature, that there is only weak evidence for a typical pattern of regional sweating [91].

The resting whole-body sweat rate might be approximately 0.4 L.min^{-1} during passive heating. The forehead (0.99 mg·cm^{-2}·min^{-1}), dorsal fingers (0.62 mg·cm^{-2}·min^{-1}) and upper back (0.59 mg·cm^{-2}·min^{-1}) would display the highest sweat flows, whilst the medial thighs and anterior legs will secrete the least (both 0.12 mg·cm^{-2}· min^{-1}) [91].

In a retrospective analysis of 506 athletes absolute and relative whole-body sweating rate across a range of sports and environmental conditions was reported. The average of absolute whole-body sweating rate in 94 male youth athletes was 0.90±0.43 L·h^{-1}, and 1.42±0.72 L·h^{-1} in adult males. The corresponding value for 40 young females was 0.63±0.18 L·h^{-1}, and 1.10±0.57 L·h^{-1} for 52 adult female athletes [92]. These data indicate higher sweating rates in adults than in young people, and more sweating in men than in women.

3.5.1.1.4.3.4 Evaporation of sweat

The following description of the physical background of evaporation is based on the book by Houdas and Ring [3].

To achieve thermal balance, the body must lose heat at rate *E*, equal to the heat storage rate, *S*. This rate of energy loss, *E*, requires the production of sweat at rate m. Using the subscript "req" (required) we define the necessary rate of heat loss (E_{req}) and the corresponding sweat production rate (m_{req}).

If the body can produce sweat at the required secretion rate, the main problem of the second part of the mechanism is to determine whether this sweat can be evaporated at the skin surface into the atmosphere. This evaporation depends mainly on the characteristics of the atmosphere. The evaporation rate from a water surface is related to the difference between the water vapour pressure at this surface and the water vapour pressure of the surrounding atmosphere. The first value depends only on the temperature of that surface, because it corresponds to the saturated water vapour pressure at that temperature, $P_{ws}{}^{o}$.

Therefore,

$$E_{max} = h_e (P^o_{ws} - P_{wa}) A_e ,$$

where h_e is the coefficient of evaporation, which depends on the convection coefficient h_e according to Lewis's relationship (h_e = 2.25 h_c), and A_e is the surface area of the water.

Therefore, if the skin were a totally wet surface, like water, the maximum evaporative heat loss rate would be E. The subscript max can then be added to E to indicate that this value is the maximal theoretical rate for evaporative heat loss from the skin. It corresponds to a rate of sweat secretion m_{max} with $m_{max} = E_{max}/2.45$. This concept is called the **maximal evaporative power (capacity) of the ambience.**

Assume an atmosphere in which the water vapour pressure is P_{wa} torr in absolute value. In this atmosphere, the body has a skin temperature T_s, which varies in relation to the air velocity, v [93]. According to the other climatic and physiological conditions, the subject must evaporate heat at rate E_{req}. Therefore, the first requirement is to produce an equivalent sweat rate, m_{req}. The second is to evaporate this sweat. The characteristics of the atmosphere determine the maximal possibility for evaporation, that is, E_{max}.

If E_{req} is smaller than E_{max}, for instance if the required heat loss is 10% of E_{max}, the sweat produced can easily be evaporated and thermal balance can be achieved. Note that in such cases the sweat being totally evaporated does not appear as drops on the skin and sweating is not apparent to the observer. The evaporative rate is independent of air humidity but depends indirectly upon air velocity. Physically speaking, E cannot be defined from humidity or velocity of the air [93].

If the total required evaporative rate E_{req} is considerably greater than E_{max}, that is when m_{req} is greater than m_{max}, only one part of the sweat produced corresponding to m_{max}, can be effectively evaporated. The actual evaporative rate E is thus limited by the physical parameters and this limit closely depends upon the humidity and the velocity of air. The difference between m_{req} and m_{max} cannot be evaporated and appears then as liquid sweat flowing on the skin. Therefore, in this case heat is not removed from the body and thermal balance cannot be achieved.

If E_{req} is somewhat less than E_{max}, the problem gets more complex, because (a) the ambient conditions are not homogenous over the whole skin surface and E_{max} can vary over each skin area, and (b) the sweat rate can also vary in each skin area. Consequently, the regional E_{req} can be either greater or smaller than the local E_{max}. In the first case, the evaporative rate is dependent upon air humidity and air velocity since part of the secreted sweat cannot be evaporated. In the second case, evaporation of the total amount of secreted sweat occurs. Thus, the evaporation rate is limited by the sweat production and is not dependent upon the physical parameters. These facts explain that, for the whole body, the evaporating power necessary for thermal homeostasis is substantially lower than the maximal evaporating power of the surrounding air [93].

Therefore, at least theoretically, thermal equilibrium should be achieved as long as E_{req} is less than E_{max}, and liquid sweat might not appear until this limit. However, this is not the case in practice as liquid sweat may appear on some skin areas although thermal balance is completely achieved, or the limit of thermal homeostasis is reached far before E_{req} becomes equal to E_{max}.

Since an individual may lose more sweat than is necessary, some authors have introduced the concept of **sweating efficiency** which is defined as the ratio of the rate of sweat evaporation to the rate of sweat production.

In 1982, Shapiro et al. published a predictive equation for sweat loss considering various environmental conditions, various workloads and clothing characteristics [94]. Valid within limits between 50 to 360 $W.m^{-2}$ for required evaporation, E_{req}, and maximal evaporative capacity, E_{max}, between 20 and 525 $W.m^{-2}$ the following equation for sweat loss, m_{sw}, was provided.

$$m_{sw} = 18.7.E_{req}.(E_{max})^{-0.455}.$$

The predicted sweat loss was strongly correlated with the measured sweat loss (r = 0.94). The ratio E_{req}/m_{sw} as found to correlate with the maximal evaporative capacity, but not with metabolic heat production [94]. In 1995, the original predictive equation for sweat loss was adjusted for outdoor conditions by evaluating separately the radiative heat exchange, short-wave absorption in the body and long-wave emission from the body to the atmosphere and integrating them in the required evaporation component (E_{req}) of the model [95].

Cheuvront et al. proposed another modification of Shapiro's equation (SE), since the equation overestimates sweating rates, particularly when applied to more extreme environmental conditions and work-loads [96]. Based on whole body sweat losses measured in 101 volunteers (80 males and females) under a wide range of environmental conditions (ambient temperature, 15–46°C; water vapour pressure, 0.27– 4.45 kPa; wind speed, 0.4 –2.5 m/s), clothing, and equipment combinations and durations (2–8 h).)the following correction of SE was made.

$$SE_c = m_{sw}\,(g\cdot m^{-2}\cdot h^{-1}) = 147\cdot \exp^{0.0012\,SE},$$

where SE is the uncorrected output from the original Shapiro equation. Applying fuzzy piecewise regression analysis to this data set, a new equation (PW) was developed:

$$PW = m_{sw}\,(g.m^{-2}\cdot h^{-1}) = 147 + 1.527\cdot E_{req}\cdot 0.87\cdot E_{max}$$

The corrected Shapiro equation and the new piecewise equation were more accurate predictors of sweating rates (58% and 65% more accurate than the original SE) and produced only minimal error for conditions both within and outside the original SE domain of validity.

Clothing layers modify the heat dissipation from the skin and affects particularly sweat cooling of the skin. Havenith and co-workers studied the effective cooling power of moisture evaporation by means of a thermal manikin [97]. They developed an equation for calculation of the effective latent heat of evaporation (λ_{eff}). The amount of latent heat varies in dependence to the distance to the skin, number of clothing layers and their permeability for water. For evaporation from the skin, λ_{eff} is close to value of water (2,430 J/g), but starts to decrease when more clothing is worn, e.g. by 11% for underwear and permeable coverall. When evaporation is from the underwear, λ_{eff} reduction is 28% wearing a permeable outer. When evaporation is from the outermost layer only, the reduction exceeds 62% (no base layer), increasing toward 80% with more layers between skin and wet outerwear. In semi- and impermeable outerwear, the added effect of condensation in the clothing opposes this effect.

3.5.1.2 Hyperthermia

In case that the heat gained cannot be dissipated, it becomes stored inside of the body and the intensity measure of heat, i.e. temperature is increasing. The "Glossary of Terms for Thermal Physiology" [2] defines **hyperthermia** as the condition of a temperature regulator when core temperature is above its range specified for the normal active state of the species. Hyperthermia may be regulated (e.g. in fever or during dehydration) or may be forced if the sum of internal (due to strenuous physical exercise) and external (environmental) heat load exceeds the capacity for heat loss.

Induced hyperthermia is the state of hyperthermia produced purposefully by increase in heat load and/or inactivation of heat dissipation by physical and/or pharmacological means. Induced hyperthermia is used in a few specific conditions as a therapeutic adjuvant. It may also be locally induced (e.g. by microwave irradiation or organ perfusion). Induced hyperthermia should be distinguished from fever induced by administration of pyrogens, which is an obsolete technique to produce elevations of body temperature in medicine for therapeutic purposes [2].

Exposure to heat may lead to several clinical pictures. The mildest of those is called **heat stress,** which is the perceived discomfort and physiological strain associated with exposure to a hot environment, especially during physical work. The next level, **heat exhaustion**, is a mild-to-moderate illness due to water or salt depletion that results from exposure to high environmental heat or strenuous physical exercise. Signs and symptoms include intense thirst, weakness, discomfort, anxiety, dizziness, fainting, and headache. Core temperature may be normal, below normal, or slightly elevated (>37°C but <40°C) [98].

Prolonged failure of heat defence mechanisms leads to **heat stroke**, which is a severe illness characterized by a core temperature

>40°C and central nervous system abnormalities such as delirium, convulsions, or coma resulting from exposure to environmental heat (classic heat stroke) or strenuous physical exercise (exertional heat stroke).

Hyperthermia, malignant: A pharmacogenetic myopathy characterized by a rapid rise in body temperature in genetically susceptible subjects during anaesthesia (triggering agents: halothane, succinyl choline, etc.). **Malignant Hyperthermia** stems from impairment (genetically originated) in the sarcoplasmic reticulum, leading to alteration in handling of the excess cytosolic Ca^{++}, skeletal muscle rigidity and spasticity and, in-turn, increased **heat production**. The often, fatal **hyperthermia** is the result of heat production in skeletal muscle and supported by the limitations of active heat dissipation during anaesthesia. It is complicated by sympathetic activation, excessive lactate production and acidosis, muscle rigidity, myoglobinaemia, and disturbances of cellular permeability [2].

Fever

The longest recognized health condition that is characterized by regulated hyperthermia is **fever.** Regulated hyperthermia means that the set point shifts upwards and all responses of thermo-effectors develop in relation to the new situation. In the initial phase of fever when the set point moves upwards, humans start shivering. The phase of fever resolution is characterized by sweating since the set point returns to its non-febrile level.

The shift of the set point may be caused by endogenic or exogenic substances, metabolites or toxins leading to fever in various conditions.

The "Glossary of Terms for Thermal Physiology" [2] defines fever as follows:

Fever is a state of elevated core temperature which is often, but not necessarily, part of the defensive responses of organisms (host) to the invasion by live (microorganisms) or inanimate matter recognized as pathogenic or alien by the host. The rise in core temperature is usually designated as due to a change in the thermo-controller characteristics resulting in an elevation of the set point of body temperature. The higher temperature is actively established and defended by the operation of heat-producing and heat-conserving thermo-effectors. Time courses and extent of natural fevers are variable, but an upper temperature limit (41°C in humans) is seldom exceeded, at which level core temperature is maintained for some time and eventually reduced when the set point of temperature regulation returns to its normal level. While any rise in body temperature may be due to fever, those rises that are not accompanied by supportive changes in thermo-effector activities, are termed hyperthermia. The humoral mechanisms underlying fever consist of the alteration of the neural control of body temperature by agents named pyrogens. Endogenous pyrogens (e.g. prostaglandin E2) are usually the obligatory mediators of fever, but direct actions of exogenous pyrogens are not ruled out. Fever can also result as an idiopathic condition of the body. It can also be caused by emotional stress. The exact mechanisms of fever induction are not yet clear.

Typical natural fever in response to bacterial infection is induced by interleukin-1 (IL-1), IL-6 and other cytokines that are formed as mediators in the multifactorial humoral host-defence response termed acute-phase response. Limited febrile rises in core temperature are apparently harmless, if bodily functions are not compromised by other disturbances or stressful conditions. Fever may contribute to the beneficial effects of the humoral host-defence responses to infection.

Romanovsky et al. proposed a different definition of fever based on the balance between ambient and deep body temperatures in independent feedback loops, avoiding a single and unique set point [99].

The influence of increased body temperature on cells and cytokines involved in the acute-phase response is not yet fully understood [100]. However, this interaction is not a matter of temperature regulation.

3.5.2 Defence against Heat Loss

Like in the protective mechanisms against heat gain, behavioural and autonomic responses can be differentiated. All mechanisms lead to less heat exchange with the environment by decreasing the area of heat dissipation and increasing the insulating power of the skin. Finally, heat is generated by initiation of non-shivering and shivering thermogenesis.

3.5.2.1 Decrease of the area of heat dissipation

Heat transfer is dependent on the area available for conductive, convective or radiative heat exchange. According to studies from the 1950s, a squatting position (Table 3.9) reduces the ratio of radiating area to total surface area significantly. Although the influence of body position on heat transfer is well established [101], the contribution of the reduction of body surface on thermal comfort and the amount of preserved heat by this behaviour remain unknown. The interaction between perceived thermal comfort and the partially autonomic responses such as changing body posture or warmth seeking behaviour, and their underlying neural pathways are topics of current research [99].

3.5.2.2 Increase of thermal insulation at the body surface

An effective barrier for heat loss can be achieved by behaviour or autonomic control processes.

3.5.2.2.1 *Behavioural effects*

Compared to the furs of arctic animals with insulation in a range of 2 to 8 clo [102], the insulation of non-glabrous skin is negligible. Wearing clothes, particularly heavily insulating hand and footwear might have been the prerequisite that allowed early mankind to migrate into cool and cold climatic zones [103].

Clo is a unit to express the relative thermal insulation values of various clothing assemblies. 1 clo = $0.18°C·m^2h·kcal^{-1}$ = $0.155°C$ $m^2·W^{-1}$. The clo was developed to express thermal insulation or thermal resistance, defined as reciprocal thermal conductivity, in practical terms. It represents the insulation provided by the normal indoor clothing of a sedentary worker in comfortable indoor surroundings. The term is used in heating and ventilation engineering in the determination of environmental conditions for human comfort [2].

The various effects of clothing on heat transfer are described in Sections 3.3.1.3 and 3.5.1.2.1.3.4. Seeking shelters and the use external heat sources complete the behavioural coping with exposure to cold.

3.5.2.2.2 *Autonomic*

Like in the exposure to heat, the mechanisms involved in the autonomic response to cold are different in whole body cooling and local cooling [73]. In either exposure of skin to coldness leads to vasoconstriction of the skin vessels, which results in an increased insulating power of the skin due to reduction of blood volume in the capillaries situated in the papillary region of the dermis.

3.5.2.2.2.1 *Response to cooling*

The reflex effect of lowering the temperature over a large skin area and the local decrease in temperature at the neuro–effector junction and the cutaneous vascular smooth muscle trigger a vasoconstrictor response in the papillary region of the skin, which leads rapidly to a graded change in skin blood flow. The reflex portion of this response is based on information arising from cutaneous cold receptors, communicated to the central nervous system and stimulation of the cutaneous vasoconstrictor nerve to increase its activity. Modest reductions of skin temperature to the range of 34°C to 32°C cause a marked reduction in skin blood flow and a concomitant increase in core temperature by 0.1°C to 0.2°C. In that range of temperatures, the reduced local skin temperature does not add much to the overall vasoconstrictor response. The reflex effects of whole body cooling affect the threshold of the core temperature for the onset of active cutaneous vasodilation, whereas local skin cooling modifies the relationship between skin blood flow and core temperature through noradrenergic activity [73].

It is no longer in question that the reflex cutaneous vasoconstriction is of sympathetic origin. However, the specific neurotransmitter(s) involved have only been identified in the recent 10 to 15 years. The predominant class of alpha adrenoceptors in the skin is the $alpha_2$ variety. The subtype-2C of alpha adrenoceptors has little or no expression on the cell membrane under warm conditions but is translocated from intracellular organelles to the cell membrane due to the action of rho kinase that is stimulated by cooling. The cascade of intra-cellular events initiated by cooling is not yet clear. Interestingly, a persistent increase in $alpha_{2C}$ adrenoceptor mediated vasoconstriction was observed after a single exposure to vibration. This may indicate a common cause of vasomotor response to cold and vibration exposure [104].

In whole body cooling but not in local cooling, the neuropeptide Y (NPY) is involved and contributes to 40% of vasoconstriction whereas 60% are due to the action of norepinephrine. A possible role for ATP as co-transmitter remains unclear.

At endothelial cells, a decrease in local temperature inhibits the endothelial nitric oxide synthase (eNOS). Consequently, production of nitric oxide (NO) is interrupted and vasodilation due to basal NO concentrations is eliminated. Figure 3.12 shows the current theory of interaction between neural activity and intercellular processes that lead to cold-induced vasoconstriction.

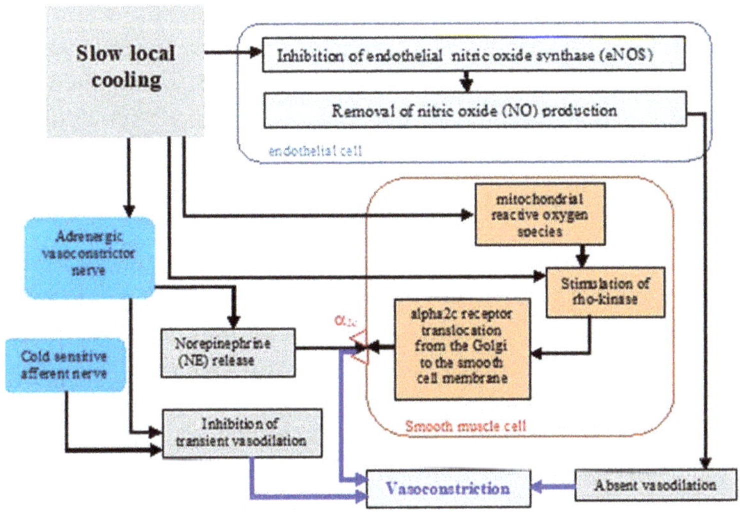

Figure 3.12 Neural and molecular mechanisms of vasoconstriction induced by local cooling. Modified from 72.

3.5.2.2.2.2 *Vasomotor mechanisms of local cooling*

Vasoconstriction leads to a reduction in blood flow rate. However, the relationship between reduction in temperature and perfusion is not linear and depends on local temperature, mean skin temperature and core temperature [103]. It has been estimated that the minimal blood flow to support cutaneous tissue is approximately 0.8 mL 100 mL^{-1} min^{-1}. A decrease of blood flow beyond nutritional needs is the basic mechanism of cold-induced injuries such as frostbites.

Glabrous and non-glabrous skin differ slightly in their response to local cooling. Whilst in hairy skin vasoconstriction lasts if the stimulus continues, in skin without hairs, for example in the fingers, the initial vasoconstriction is reversed after a short time and skin blood flow start to fluctuate up and down. This phenomenon was called "hunting reaction" or "cold-induced vasodilation", which was first described in 1930 by T. Lewis, who hypothesized an axon reflex as primary cause of cold vasodilation. The repeated alterations in skin perfusion are explained by active involvement of arterio-venous anastomoses (AVAs), which becomes more apparent when body core temperature is undisturbed, the skin is warm, the medium temperature for local cooling is very low, at low altitudes and absent mental stress. The hunting reaction may be modified by acclimatization to cold climate [105]. Cold-induced vasodilation was also observed in glabrous skin but without the typical fluctuations in blood flow [73].

3.5.2.3 Internal heat production

Non-shivering thermogenesis is the first response to cold exposure when vasoconstriction reached the limit of further increase of insulation and thereby minimizing heat loss. Simultaneously, other mechanisms of facultative thermogenesis are initiated such as heat production due to non-exercise activity and cold-induced shivering in response to a cold environment. However, the contribution of voluntary non-exercise activity to cold-induced thermogenesis remains unclear [106].

Both the threshold for cold-induced thermogenesis and for sweating define the thermo-effector threshold (inter-threshold) zone, in which other than vascular thermoregulatory responses are evoked. The threshold for adaptive thermogenesis is approximately 2°C below thermoneutral temperature depending on the individual history of acclimatization to cold. It is well known, that some volatile and intravenous anaesthetics affect thermoregulation by altering threshold values for cold-induced vasoconstriction and shivering, promoting in this way perioperative hypothermia [107].

3.5.2.3.1 *Involuntary heat production*

Input from thermoreceptors in the skin, spinal cord or brain can induce non-shivering and shivering thermogenesis. Responding neurons originating from the hypothalamus and the pre-optic area, stimulate brown fat cells via beta-adrenergic receptors leading to an immediate increase in activity of UCP-1 and to lipolysis. UCP-1 gene transcription, mitochondrial biogenesis, hyperplasia of brown adipose tissue and recruitment of brown adipocytes in white adipose tissue depots is established under chronic stimulation [108]. Detailed descriptions of central neural pathways involved in cold defence have been published [109, 110].

3.5.2.3.1.1 *Non-shivering thermogenesis*

The sympathetically controlled brown fat is still in focus of research on non-shivering thermogenesis. The question whether other tissues than brown fat are involved in non-shivering thermogenesis, remains unanswered, although several candidates such as skeletal muscle [57, 108, 112,] liver [108] and white adipose tissue [108] have been proposed.

Thyroid hormones have an important influence on both obligatory and adaptive thermogenesis. Their major thermogenic mechanisms include an increased turnover of adenosine triphosphate (ATP) and a programmed reduction in the efficiency of ATP synthesis. Consequently, high activity of the thyroid gland such as in hyperthyroidism leads to an increase in metabolic rate and an acceleration of synthetic and catabolic pathways for all three macronutrients, namely proteins, carbohydrates, and lipids. Hypothyroidism is associated with a reduction of basal metabolic rate up to 30% [111].

Thyroid hormones affect the adaptive thermogenesis in a very complex interaction with sympathetically released norepinephrine, which is directed to the activity of uncoupling protein-1(UCP-1) in brown fat. Hypothyroidism results in a reduced rate of facultative heat production despite a high sympathetic activity [112].

3.5.2.3.1.2 *Shivering thermogenesis*

Shivering as an effective way of heat production. Maximal shivering can increase metabolic heat production to the three or six-fold of the heat produced at rest and can be continued at high level for 3 to 4 [107]. Typically, cold-induced shivering begins in the muscles of the upper part of the body, including the masseters (the classical chattering of the teeth) and particularly the muscles of the shoulder girdle. From there, it spreads to the paravertebral muscles and to the lower limbs [3].

Shivering thermogenesis is regulated by feed forward mechanisms activated by peripheral thermoreceptors. During cold exposure, peripheral thermal afferents reach the pre-optic area of the hypothalamus. Neural circuits between dorsomedial hypothalamus and the rostral medullary raphe lead finally to activation of γ-motor neurons that innervate muscle spindles, which results in an increase in muscle tone and the onset of shivering [112].

Electromyography (EMG) is the recommended method for the detection of shivering. The frequency of shivering bursts is about 8–10 Hz; however, also continuous shivering at a frequency 0.1–0.2 Hz was identified in the EMG. The intensity of shivering varies between 2–5% and 7–15% of maximal voluntary contraction (%MVC) and varies greatly amongst muscles and individuals. While the burst shivering rates are similar in women and men, women display a higher shivering intensity than men at the same given metabolic rate in the cold [113].

The observation of obvious shivering is an important clinical sign for grading hypothermia. Shivering may be maximal at 32°C to 33°C, may continue at 31°C, and may not cease until core temperature drops to approximately 30°C [114].

Based on clinical observation the Bedside Shivering Assessment Scale was developed as a simple tool evaluating the metabolic stress of shivering. This was evaluated against indirect calorimetry in 50 cerebrovascular patients after therapeutic temperature modulation and gained a high level of inter-observer reliability [115]. Good reliability was also confirmed amongst diverse care practitioners [116].

As in other forms of thermogenesis, energy, i.e. ATP, is required for maintaining shivering which is obtained through the combined oxidation of carbohydrates, lipids and proteins. The relative contribution of carbohydrates and lipid to the total of produced heat vary in dependence to the intensity of shivering and available carbohydrate reserves. At low intensities of shivering equal to 2 resting metabolic rates (RMR), carbohydrates contribute to 30%

of total heat; however, at shivering intensities of 6 RMRs, 80% of heat is due to oxidation of carbohydrates. The pattern of fuel selection during cold-induced shivering is definite different from that observed during exercise. The large difference in metabolic rate (1 to 5× RMR for shivering vs. 1 to 20× RMR for exercise) is reflected in oxidation rates that are 2 to 7 times lower for shivering than for exercise [113].

Several drugs and non- pharmacologic interventions inhibit cold-induced shivering by lowering the threshold. The pharmacologic attenuation of shivering can be induced by anaesthetics such as halothane, isoflurane or enflurane, opioids such as meperidine and high dosages of morphine, alfentanil or fentanyl. Inhibition of shivering can be achieved by the application of the α2-agonistic substance clonidine. Serotonin uptake inhibitors such as tramadol and serotonin agonists/antagonists such as ketanserin, buspirone or ondansetron lower the threshold for shivering. Cholinomimetics such as physostigmine and *N*-methyl-D-aspartate (NMDA) antagonists such as magnesium sulphate have been successfully used to reduce shivering. Other pharmaceutical agents have been used or are under evaluation for the inhibition of shivering [117].

Non-pharmacologic treatments can also modify cold-induced shivering. Breath holding, muscle relaxation and mental arithmetic reduced EMG registered shivering by 30%. Upright posture and exercising in cold environment attenuates shivering. The effect of local skin warming remains unclear, although an increase of 4°C in mean skin temperature results in a 1°C decrease in shivering threshold [117].

3.5.2.4 Hypothermia

The "Glossary of Terms for Thermal Physiology" defines hypothermia [2] as the condition of a temperature regulator when core temperature is below its range specified for the normal active state of the species. (Greek: *hypo*-beneath; *therme*-heat).

In several birds and mammals, hypothermia may be *regulated* (e.g. torpor, hibernation). Apart from bears, hibernating animals are generally small in body size and include echidna, marsupial, insectivora such as the European hedgehog, bats and rodents [118]. In humans, only forced hypothermia occurs when heat loss exceeds the capacity for total heat production. Forced hypothermia can be further differentiated into accidental hypothermia and induced hyperthermia for therapeutic purpose.

3.5.2.4.1 *Accidental hypothermia*

Accidental hypothermia is the condition of a temperature regulator following an accidental or deliberate decrease in core temperature below its range specified for the normal active state of the species. The condition usually occurs in a cold environment and is associated with an acute problem, but without primary pathology of the temperature regulating system [2].

Accidental hypothermia is a condition associated with significant morbidity and mortality [119]. Primary hypothermia occurs when heat production in an otherwise healthy person is less than the heat loss to an excessive cold environment, especially when the energy stores of the body are depleted. Depending on clothing and the ambient medium, negative heat storage occurs and increases with time, while internal temperature progressively falls.

Accidental hypothermia caused by environmental exposure can occur during any season, even in temperate or tropical climates. Cold and wet environments pose the greatest risk, particularly at high latitudes, in subarctic regions or after long immersion in water with temperature below 20°C. Accidental hypothermia can occur in people participating in outdoor work or recreation, including wilderness travellers. Hypothermia is associated with urban homelessness and with the use of alcohol [114].

Secondary hypothermia can occur in severely ill persons with a wide variety of medical conditions including intoxication, endocrinological, cardiovascular or failure of the central and/or peripheral nerve system. In all types of hypothermia, consciousness, breathing, and circulation are initially intact but are impaired as the body cools.

The standard classification of hypothermia is based on core temperature with the classes mild: 35°C to 32°C, moderate: 32°C to 28°C, and severe/profound: <28°C. Outside hospital when core temperature measurement may be difficult, the Swiss hypothermia classification system [120] is recommended for staging (Table 3.15) since it allows estimation of core temperature based on clinical signs.

Table 3.15 Swiss hypothermia (HT) staging system [119]

Stage	Clinical signs	Estimated core temperature
Hypothermia I (mild)	Clear consciousness with shivering	35°C to 32°C
Hypothermia II (moderate)	Impaired consciousness, may or may not be shivering	<32°C to 28°C
Hypothermia III (severe)	Unconscious; vital signs present	<28 to 24°C
Hypothermia IV (severe)	Apparent death, vital signs absent	<24°C

Figure 3.12 illustrates the development of clinical signs in accidental hypothermia in dependence to rectal temperature and heat loss. Although Fig. 3.13 is based on data published before 1978, it describes a predictable pattern of organ dysfunction and associated clinical manifestations in progressive hypothermia [121].

In mild hypothermia with a core temperature between 32°C and 35°C, confusion, amnesia, dysarthria and ataxia are observed due to reduced cerebral metabolism. At this stage the endocrine and metabolic function is characterized by shivering, hyperglycaemia, increase in catecholamines and oxygen consumption. The cardiovascular system responds with an increase in heart rate, cardiac output and blood pressure [121]. Prolonged PR and QT intervals may appear in electrocardiograms. The initial response in the respiratory system to mild hypothermia is an increase of both respiratory rate and minute ventilation. In addition, an excessive amount of watery mucus is produced. Mild or moderate hypothermia affect fluid and water balance by reduction of plasma volume and induction of diuresis that can increase fluid loss almost two-fold. However, the mechanism of cold-induced diuresis remains unclear [123].

Core temperatures between 28°C and 32°C define the stage of moderate hypothermia. In this phase all body functions slowly decline [121]. The level of consciousness becomes progressively reduced, hallucinations may appear, and the pupils dilate. Shivering ceases and metabolism and oxygen consumption is diminishing. Heart rate and cardiac output decline progressively. J-waves, also named Osborn waves are detected in about 80% patients with a core temperature less than 32.2°C [122]. Atrial fibrillation is common when the core temperature is less than 32°C and is not worrisome in the absence of other signs of cardiac instability [119]. However, the risk of ventricular arrhythmia is growing. Airway protection is lost while respiration rate and minute ventilation decline progressively. Haematocrit rises 2% per 1°C decline in core temperature and coagulopathy may develop.

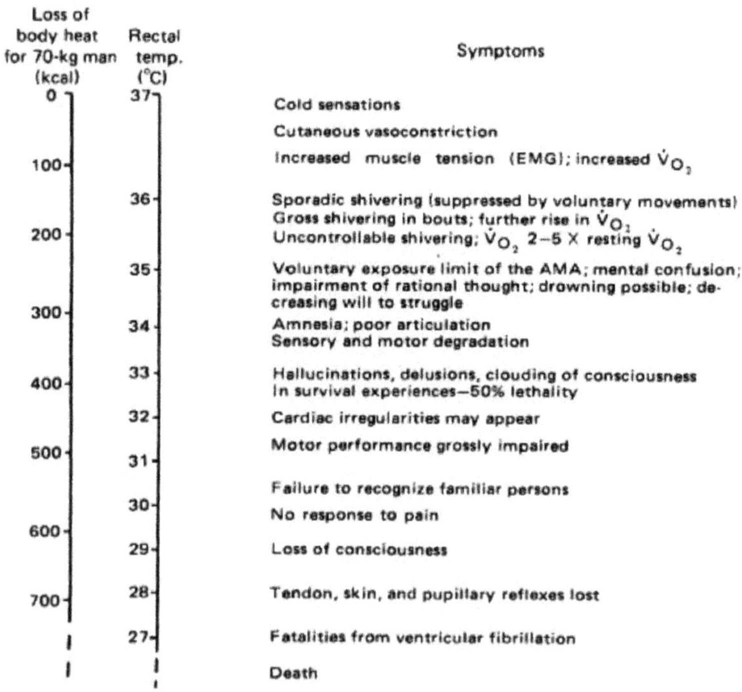

Figure 3.13 The progressive failure of body systems owing to heat loss. Reproduced from [3], with kind permission from Springer Science+Business Media B.V.

In severe hypothermia with a core temperature less than 28°C, organ function is close to failure. The patient is unconsciousness, there is a global loss of reflexes, the electroencephalographic (EEG) activity decreases constantly and EEG silence occurs at temperatures below 26°C. Blood pressure, heart rate and cardiac output are diminished, ventricular arrythmia is common and asystole occurs at 20°C. Apnoea starts to develop at temperature less than 24°C and pulmonary oedema may manifest. Reduction in renal perfusion leads to oliguria [123].

Standards for diagnosis, staging, triage and treatment of hypothermia may increase the survival chance of patients with severe hypothermia. A recent update review on accidental hypothermia includes a list of the most extreme reported accidental hypothermia cases including the case of a 29-year-old woman who survived and fully recovered from a body core temperature of 13.7°C [124]. Long-time survival in 33% of severe accidental hypothermia cases with cardiac arrest was reported. Fifty-three percent of survivors presented with normal neurological outcome, 13% showed clear neurological impairment and the remaining were classified as borderline [125]. A recent retrospective study reported a survival rate of 37% based on a sample of 286 hypothermic cardiac arrest patients [126].

There is a long tradition to predict survival time from water temperature and the duration of cold immersion. While the early models relied only on water temperature or linear extrapolation of heat loss, modern complex models consider thermal physiology and several anthropometric and ambient conditions. Based on the body heat balance equation with consideration of the contribution of core temperature dependent heat production through shivering, the Cold Exposure Survival Model (CESM) predicts the time needed to establish profound hypothermia [127]. This model can predict within limits survival time for cold exposure in either air or water. However, it should be used as a guideline what can be expected and not as reference values. Windy weather and turbulent sea lead to overestimation of predicted survival times [128].

Primary accidental hypothermia can result from cold-water immersion [129], cold-water submersion [129] and exposure to other environment such as such as cold air [129], burial in avalanches [130] or captured in crevasses [130]. Cold-water immersion is the predominant approach to investigate the physiological responses to cold. Four stages of cold immersion have been described associated with particular risk for fatal outcome. These are: (1) initial responses to immersion (0–3 min), (2) short-term immersion (3–15 min), (3) long-term immersion (> 30 min) and (4) post immersion [131].

Normally, the possibility of hypothermia does not occur within 30 min. However, the initial phases of cold-water immersion may be also life threatening.

Tipton was the first who used the term "cold shock" for the description of the stimulating, emotive aspect of the response that occurs within the first 3–4 min of cold-water (head-out) immersion [131]. The cold shock results in a range of responses such as an initial peripheral vasoconstriction, the gasp reflex, hyperventilation and tachycardia which can, respectively, lead to hypocapnia, the inability to breath-hold, hypertension, and increased cardiac output. The initial cardiovascular responses to cold-water immersion represent only a slight risk for healthy individuals. However, in individuals with pre-existing cardiovascular or respiratory disease, the cold shock responses may cause sudden death either immediately or within a matter of minutes after immersion due to syncope or convulsions leading to drowning, or vagal arrest of the heart, and ventricular fibrillation [129].

The most obvious physical effect of short-term immersion is an impaired function of the skeletal muscles. Decreased nerve conduction velocity, slow rates of diffusion, increased viscosity result from rapid limb cooling leading to fatigue and reduced muscle force that impair swimming performance. Swimming failure contributes significantly to drowning [132].

The last phase of cold immersion deserves special attention since approximately 17% of immersion deaths occur just before, during or immediately following rescue [132]. The mechanism causing pre-rescue collapse remains unclear. A possible cause might be a reduction in catecholamine secretion and the consequent withdrawal of their protective effect. Collapse during rescue was related to redistribution of blood volume, hypovolaemia, increased blood viscosity and impaired perfusion of the heart and skeletal muscles. Due to the influence of gravity on circulation, vertical body position of the casualty should be avoided [132].

The most important cause of post-rescue is hypoxia, secondary to the aspiration of water. Inappropriate rewarming is a less common cause of post-rescue death. Aggressive external rewarming of hypothermic subjects leads to peripheral vasodilation that demands compensation before the body has regained the blood volume necessary for that. Such a failure of blood redistribution may result in collapse of arterial blood pressure and poor coronary perfusion [132].

Historically, post-rescue death was causative related to the after-drop of core temperature. At the end of the 18th century, James Currie was probably the first to notice the "after drop" in body temperature which occurs when cooling is stopped, and the further fall in body temperature which happens when a peripheral vasodilatation is evoked in a chilled man [133].

The theoretical risk is that after-drop will either inhibit shivering thermogenesis or, more importantly, put the patient into the temperature range where spontaneous malignant cardiac arrhythmias are possible. Convective and conductive mechanisms have been proposed to explain after-drop. The convective theory relates after-drop to the return of cold blood from the periphery to the core while the conductive theory is one of simple temperature equilibration between the cold periphery and the relatively warmer core. In published trials on rewarming hypothermic patients, the after-drop in core temperature is typically <0.5°C and is least with active core rewarming [134]. The incidence of severe after-drop post-rescue or after rewarming remains unknown. Consequently, the risk for a fatal of outcome of severe after-drop is undefined.

3.5.2.4.2 *Induced hypothermia*

Induced Hypothermia is the state of hypothermia produced purposefully by increasing heat loss from the body and/or inactivation of heat conservation and heat production by physical and/or pharmacological means [2].

Historically, hypothermia was induced prior to surgery to enable procedures with prolonged ischemia, such as open-heart surgery and organ transplant. The protective effect of hypothermia is explained by the reduction of cellular metabolism and oxygen demand, leading to modification of a pathological process caused by acute reintroduction of oxygenated blood following a period of ischaemia. Two windows of opportunity for therapeutic hypothermia exist for these ischaemia/reperfusion injuries. The peak of free radical production in the short phase of initial reperfusion defines the first window after ischaemia. Treatment during this period likely requires cooling during ischaemia or cardiac arrest so that the blood is already cooled at reperfusion. The other opportunity lasts several hours, and cooling is intended to attenuate apoptosis and cell necrosis after blood flow has been restored [135].

Although practised since the 1950s, therapeutic hypothermia is not an established treatment option. The cardiopulmonary bypass (CPB) circuit is integral to most cardiac surgeries. However, the most appropriate temperature for CPB is still controversial since a favourable outcome of cold versus warm or normothermic CPB is not yet evidenced [136].

Since a variety of temperature targets are now used, the 2015 updated American Heart Association Guidelines for Cardiopulmonary Resuscitation and Emergency Cardiovascular Care adopted the term **targeted temperature management** to refer to induced hypothermia as well as to active control of temperature at any target [137]. The following three recommendations are made:

1. It is recommended that comatose (i.e. lack of meaningful response to verbal commands) adult patients with return of spontaneous circulation after cardiac arrest have targeted temperature management. This applies to the subgroup of patients with out-of-hospital cardiac arrest due to ventricular fibrillation/pulse less ventricular tachycardia and to patients within-hospital cardiac arrest caused non-ventricular fibrillation/ pulse less ventricular tachycardia.
2. The guideline recommends selecting and maintaining a constant temperature between 32°C and 36°C during targeted temperature management.
3. The third recommendation is **against** the routine pre-hospital cooling of patients after return of spontaneous circulation with rapid infusion of cold intravenous fluids.

The neuroprotective potential of hypothermic treatment was often claimed, but the results from systematic reviews and studies are contradictive. The guidelines for "Early Management of Patients with Ischemic Stroke" state that insufficient clinical evidence is the main obstacle to establish a class of recommendation for induced hypothermia in acute stroke. Consequently, no recommendations were made, either for or against hypothermic treatment [138].

Newly born infants born at term or near-term with evolving moderate to severe hypoxic-ischemic encephalopathy should, where possible, be offered therapeutic hypothermia. This treatment should be started within 6 h of birth, continue for 72 h of birth and re-warm over at least 4 h [139]. However, concerns were raised about the safety of induced hypothermia in neonatal infants, since an increased risk of thrombocytopenia and cardiac arrhythmia during intervention was reported [140].

Other applications of targeted temperature management such as traumatic spinal cord injury [141], traumatic brain injury [142] or safety issues in patients with coagulation disorders are under clinical investigation [143], but final conclusions on effectiveness cannot be drawn yet.

3.6 Adaptation

The environmental conditions on earth vary in respect to time and location. Living organisms developed adaptive mechanisms to cope with and survive in such an environment. Adaptation includes genetic, physiologic, morphological or behavioural responses.

The "Glossary of Terms for Thermal Physiology" defines **adaptation** as changes that reduce the physiological strain produced by stressful components of the total environment. This change may occur within the lifetime of an organism (phenotypic) or be the result of genetic selection in a species or subspecies (genotypic). Acclimation as defined in [2] relates to phenotypic adaptations to specified climatic components. In thermal physiology, the use of the term adaptation does not require specification of the climatic components of the total environment to which the organism adapts, but the most obvious component is often denoted (e.g. adaptation to heat). There are no distinct terms that relate genotypic adaptations to the climate or components of climate [2]. In comparison to adaptation as defined in neurophysiology, the adaptive processes in thermal physiology usually occur with larger time constants.

In humans, it is an unproven hypothesis that ethnic differences exist other than those associated with phenotypic thermal adaptation. Most of research in the past studied indigenous populations in their traditional habitats and induced thereby confounding influences. Differences in morphological configuration and their impact on heat exchange with the environment must be considered because they may be misinterpreted as genetic influence of thermal adaptation [144].

Physiological adaption should not be confused with adaptations of systems as defined in the glossary for biometeorology [145]. Adaptation in human systems is defined by the Intergovernmental Panel on Climate Change (IPCC) as "The process of adjustment to actual or expected climate and its effects, in order to moderate harm or exploit beneficial opportunities". In natural systems, the IPCC defines adaptation as the process of adjustment to actual climate and its effects, and notes that human intervention may facilitate adjustment to the expected climate.

Acclimation is defined by physiological or behavioural changes occurring within an organism, which reduces the strain or enhances endurance of strain caused by experimentally induced stressful changes in particular climatic factors [2].

The terms acclimation and acclimatization are etymologically indistinguishable. Both words have been assigned several and different meanings [146]. The most useful of the assigned meanings, adopted here, would seem to be those of Hart [147] and Eagan [148] who use the term **acclimation** to describe the adaptive changes that occur within an organism in response to *experimentally induced changes in particular climatic factors* such as ambient temperature in a controlled environment, and the term **acclimatization** to describe the adaptive changes that occur within an organism in response *to changes in the natural climate* [2].Consequently, **acclimatization** was defined as the process by which a living organism becomes adapted to a change of climatic environment [145]. Acclimatization is a relatively quick adaptation that occurs within the organism's lifetime [2]. Traditionally, acclimatization studies in humans have focused on adapting to temperature changes (such as moving from a temperate climate to the tropics), but acclimatization can also occur with a change in altitude, photoperiod and humidity, for instance [145].

Adaptations can be purposefully induced by challenging the physiological systems with regularly repeated stimuli that are strong enough to provoke a response. This is the same principle as that is used in physical training for enhancing muscle force and/ or endurance. However, repeated stimulation leads to **habituation**, which is a reduction of responses to or perception of regular stimuli [2]. Acclimation is transient and gradually disappears if not maintained by continued repeated exposure.

Since temperature acclimation evokes a complex multi-system response, changes are expected in temperature regulation as well as in the associated systems that control cardiovascular, endocrine and nervous functions. Research on the human body's ability to adapt to thermal environment require a clear and detailed definition of the aim of the study, consideration of the functional capacity and anthropometric characteristics of individual participants, definition of intensity, duration, frequency, and number of thermal exposures. Core and skin temperatures are mandatory outcome measures in combination with a main outcome of associated systems under investigation. While workers, military and athletes may benefit from thermal adaptation, the effectiveness of planned adaptation to the thermal environment for reducing temperature related mortality and illness remains inconclusive [149].

3.6.1 Adaptation to Heat

Most of the available research is related to heat adaptation. Acclimation regimens of different duration have been developed, ranging from short-term (<7 days) over medium (8–14 days) to long-term (>15 days). Heat load is generated by either a constant work-rate or self-regulated exercise or through controlled hyperthermia. The latter was also named isothermal heat acclimation and is achieved through exercise in combination with external heating [150]. Passive heating through immersion in water of 40 to 42°C or sauna bathing in dry air of 70–90°C and recently controlled exercise intensity were used for heat acclimation [151].

Classic signs of heat acclimation include a cooler body temperature (core and skin) and enhancement in the sweat response after short-term acclimation. Long-term acclimation results in sudomotor habituation and improved sweat efficiency, with a reduction in sweat secretion and drippage [150]. Heat acclimation leads to a decreased threshold for cutaneous vasodilation resulting in increased skin blood flow rate. Typical cardiovascular responses are a lowered heart rate; a better sustained cardiac output and better defended blood pressure and an increase in stroke volume, myocardial compliance and efficiency. Skeletal muscle metabolism is improved, muscle glycogen is spared, the lactate threshold is raised while muscle and plasma lactate are lowered, and muscle force production is enhanced [152].

Adaptations related to the induction of heat acclimation are typically evaluated in relation to the attainment of a plateau in responses such as plasma volume, heart rate, exercise capacity, thermal comfort, sweating rate, mean skin temperature and core temperature [151]. The decay in heat acclimation can be estimated by using the following formula:

$$\%\,\text{decay} = \frac{\text{Acclimatized value} - \text{Value after decay}}{\text{Acclimatized value} - \text{Unacclimatized value}}$$

However, heat acclimation can be re-induced through additional bouts of heat training. The number of days to reach prior full acclimation status can be used for the quantification of re-acclimation [151]. A recent systematic review concluded that five or more heat acclimation days induce stable adaptations in end-exercise heart rate and core temperature, but for every day without heat exposure (i.e. decay day) 2.5% of these adaptations are lost. Extending daily heat exposure duration improves adaptations in core temperature. For sweat rate, longer heat acclimation periods are related to more sustainable adaptations [151].

Thermo tolerance or acquired cellular thermo tolerance describes the cellular adaptation accompanying systemic changes induced by successful heat adaptation [153]. This **r**apid, short acting molecular process is associated with the synthesis of several families of heat shock proteins (HSP) of different molecular weights resulting from acute short sub-lethal heat injury. It is thought to protect cells from noxious stimuli as well as to accelerate their repair. It is also defined as heat shock response (HSR). The time course of heat shock proteins varies in different cells but, on the average, heat shock proteins (HSP) in the intact body seem to operate several hours following the stress and retains its activity for a few days. The response is not heat-specific and can be elicited after subjection to several other stressors (e.g. ischemia, some chemicals, etc.) [2].

3.6.2 Adaptation to Cold

Cold is an underrated risk factor for health [154] and the association between cardiovascular events such as cardiac infarction and exposure to cold in predisposed individuals is supported by epidemiological studies. Although exposure to cold evoke in humans genetic, physiologic, morphological or behavioural adaptation responses, only the latter may prevent some fatalities associated with exposure to cold.

Like in adaptation to heat, genetic adaptations to cold may have developed in some ethnicities, but conclusive evidence is not yet available [144, 156].

The main cold adaptation responses are either insulatory (circulatory adjustments) or metabolic (shivering or non-shivering thermogenesis) and may be positive (enhanced) or negative (blunted) [155]. The pattern of cold adaptation is dependent on the medium (air, water) and intensity (continuous, intermittent) and site (local, total body) of the cold stimulus. Several individual factors like age, sex, body composition, exercise, diet, fitness and health modify the responses to cold. Habituation of thermal sensations to cold develops first, followed by cardiovascular, metabolic and endocrinological responses. Like in heat acclimation, adaptation will gradually disappear if the repeated cold stimulus is discontinued.

According a model proposed by Young, the magnitude of body-heat loss is the key determinant for development of cold adaptation patterns. When repeated short cold exposures involve only limited parts of the body the resulting minor loss of body heat will promote habituation characterized by blunted shivering and blunted cutaneous vasoconstriction. A significant loss of deep body heat elicits two possible responses in dependence of whether the reduction in core temperature can be compensated by thermogenesis (either shivering or non-shivering). In case of non-compensable heat loss leading to a significant decline in deep body temperature during exposures, insulation through thermoregulatory adjustments are initiated which include enhanced cutaneous vasoconstriction and redistribution of body heat towards the subcutaneous shell increasing thereby muscle blood flow. The other possible response is metabolic acclimation if thermogenesis can sufficiently defend core temperature, and both non-shivering and shivering-induced heat production are enhanced [157].

Depending on the degree of the cold exposure, two levels of whole body cold acclimation can be distinguished [156]. Severe cold exposure results in a drop in mean skin, tissue and body core temperature, which is generally evoked by using repeated cold-water immersions leading to adaptation of insulation. In contrast, moderate cold exposure leads to decreased skin and tissue temperatures with no or minor drop in body core temperature representing metabolic adaptation that is typically the result of repeated exposure to cold air.

Based on the observation that humans residing or working in cold environments exhibit a stronger cold-induced vasodilation reaction in the peripheral microvasculature than those living in warm regions of the world, led to a general assumption that thermal responses to local cold exposure can be systematically improved by natural acclimatization or specific acclimation. However, a systematic review on the trainability of cold-induced vasodilation concluded that repeated local cold exposure does not alter circulatory dynamics in the periphery, and that extended stays in cold environments does not reduce the risk of cold injuries [157].

The functional significance of physiological cold adaptation is unclear. Behavioural and cultural responses, particularly clothing and houses with adequate heating facilities possess the highest protective potential against cold.

References

1. Patrick GL. *Introduction to Medicinal Chemistry*. 5th ed., Oxford University Press, 2013.
2. IUPS Thermal Commission. Glossary of terms for thermal physiology. *Japan Journal of Physiology* 2001, 51(2), 245–280.
3. Houdas Y, Ring EFJ. *Human Body Temperature. Its Measurement and Regulation*. Plenum Press, New York, London, 1982.
4. Imeri L. Thermoregulation as a non-unified system: A difficult to teach concept, *Temperature* 2017, 4(1), 1–8.
5. Vriens J, Nilius B, Voets T. Peripheral thermosensation in mammals. *Nature Reviews Neuroscience* 2014, 15, 573–589.

6. Goldscheider A. Temperatursinn des Menschen. In: Bethe A, Bergmann V, Emden G, Ellinger G, Hrg, A, eds, *Handbuch der normalen und pathologischen Physiologie mit Berücksichtigung der experimentellen Pharmakologie. Elfter Band: Receptionsorgane I*, Springer Verlag, Berlin-Heidelberg, 1926, pp. 131–164.
7. Vay L, Gu CJ, McNaughton PA. The thermo-TRP ion channel family: Properties and therapeutic implications. *British Journal of Pharmacology* 2012, 165, 787–801.
8. Hensel H, Iggo A, Witt I. A quantitative study of sensitive cutaneous thermoreceptors with C afferent fibres. *The Journal of Physiology* 1960, 153(1), 113–126.
9. Romanovsky AA. Thermoregulation: Some concepts have changed. Functional architecture of the thermoregulatory system. *American Journal of Physiology - Regulatory, Integrative and Comparative Physiology* 2007, 292, R37–R46.
10. Brock JA, McAllen RM. Spinal cord thermosensitivity: An afferent phenomenon? *Temperature* 2016, 3(2), 232–239.
11. Shafton AD, McAllen RM. Location of cat brain stem neurons that drive sweating. *American Journal of Physiology - Regulatory, Integrative and Comparative Physiology* 2013, 304, R804–R809.
12. Shibasaki M, Crandall CG. Mechanisms and controllers of eccrine sweating in humans. *Frontiers in Bioscience - Scholar* 2010, 2, 685–696.
13. Moran JL, Peter JV, Solomon PJ, Grealy B, Smith T, Ashforth W, Wake M, Peake SL, Peisach AR. Tympanic temperature measurements: Are they reliable in the critically ill? A clinical study of measures of agreement. *Critical Care Medicine* 2007, 35(1), 155–164.
14. Lefrant JY, Muller L, Coussaye JE, Benbabaali M, Lebris C, Zeitoun N, Mari C, Saissi G, Ripart J, Eledjam JJ. Temperature measurement in intensive care patients: Comparison of urinary bladder, oesophageal, rectal, axillary, and inguinal methods versus pulmonary artery core method. *Intensive Care Medicine* 2003, 29, 414–418.
15. Erickson RS, Kirklin SK. Comparison of ear-based, bladder, oral, and axillary methods for core temperature measurement. *Critical Care Medicine* 1993, 21(10), 1528–1534.
16. Robinson J, Charlton J, Seal R, Spady D, Joffres MR. Oesophageal, rectal, axillary, tympanic and pulmonary artery temperatures during cardiac surgery. *Canadian Journal of Anaesthesia* 1998, 45(4), 317–323.
17. Mellergard P. Intracerebral temperature in neurosurgical patients: Intracerebral temperature gradients and relationships to consciousness level. *Surgical Neurology* 1995, 43, 91–95.

18. Otawara Y, Ogasawara K, Kubo Y, Tomitsuka N, Ogawa A, Suzuki M. Brain and systemic temperature in patients with severe subarachnoid hemorrhage. *Surgical Neurology* 2003, 60, 159–164.

19. Mcilvoy L. Comparison of brain temperature to core temperature: A review of the literature. *Journal of Neuroscience Nursing* 2004, 36(1), 23–31.

20. Stelfox HT, Straus SE, Ghali WA, Conly J, Laupland K, Lewin A. Temporal artery versus bladder thermometry during adult medical-surgical intensive care monitoring: an observational study. *BMC Anesthesiology* 2010, 10(1), 13.

21. Kawanami S, Horie S, Inoue J, Yamashita M. Urine temperature as an index for the core temperature of industrial workers in hot or cold environments. *International Journal of Biometeorology* 2012, 56(6), 1025–1031.

22. Sund-Levander M, Forsberg C, Wahren LK. Normal oral, rectal, tympanic and axillary body temperature in adult men and women: A systematic literature review. *Scandinavian Journal of Caring Sciences* 2002, 16(2), 122–128.

23. Mazerolle SM, Ganio MS, Casa DJ, Vingren J, Klau J. Is oral temperature an accurate measurement of deep body temperature? A systematic review. *Journal of Athletic Training* 2011, 46(5), 566–573.

24. Gass EM, Gass GC. Rectal and oesophageal temperatures during upper and lower-body exercise. *European Journal of Applied Physiology* 1998, 78, 38–42.

25. Byrne C, Lim CL. The ingestible telemetric body core temperature sensor: A review of validity and exercise applications. 2007, 41(3), 126–133.

26. Craig JV, Lancaster GA, Williamson PR, Smyth RL. Temperature measured at the axilla compared with rectum in children and young people: Systematic review. *BMJ* 2000, 320(7243), 1174–1178.

27. Charafeddine L, Tamim H, Hassouna H, Akel R, Nabulsi M. Axillary and rectal thermometry in the newborn: Do they agree? *BMC Research Notes*, 2014, 584.

28. Craig JV, Lancaster GA, Taylor S, Williamson PR, Smyth RL. Infrared ear thermometry compared with rectal thermometry in children: A systematic review. *The Lancet* 2002, 360(9333), 603–609.

29. Daanen HAM. Infrared tympanic temperature and ear canal morphology. *Journal of Medical Engineering & Technology* 2006, 30, 224–234.

30. Allegaert K, Casteels K, Van Gorp I, Bogaert G. Tympanic, infrared skin, and temporal artery scan thermometers compared with rectal measurement in children: a real-life assessment. 2014, 76, 34–38.

31. Ring EFJ, McEvoy H, Jung A, Zuber J, Machin G. New standards for devices used for the measurement of human body temperature. *Journal of Medical Engineering & Technology* 2010, 34(4), 249–253.

32. Bitar D, Goubar A, Desenclos JC. International travels and fever screening during epidemics: A literature review on the effectiveness and potential use of non-contact infrared thermometers. *Euro Surveillance* 2009, 14(6), 1–5.

33. International Standard IEC/FDIS 80601-2-59 ISO/TC 121/SC 3. Medical electrical equipment —Part 2-59: Particular requirements for the basic safety and essential performance of screening thermographs for human febrile temperature screening, 2017.

34. Ring EFJ, Jung A, Kalicki B, Żuber J, Rustecka A, Vardasca R. New standards for fever screening with thermal imaging systems. In: Ring F, Jung A, Zuber J, eds, *Infrared Imaging: A Casebook in Clinical Medicine.* IOP Publishing, Bristol, 2015, pp. 5.1–5.10.

35. Chudecka M, Lubkowska A. Thermal maps of young women and men. *Infrared Physics & Technology* 2015, 69, 81–87.

36. Marins JCB, Moreira DG, Cano SP, Quintana MS, Soares DD, de Andrade Fernandes A, da Silva FS, Costa CMA, dos Santos Amorim PR. Time required to stabilize thermographic images at rest. *Infrared Physics & Technology* 2014, 65, 30–35.

37. Marins JCB, Fernandes AA, Cano SP, Moreira DG, da Silva FS, Costa CMA, Fernandez Cuevas I, Sillero-Quintana M. Thermal body patterns for healthy Brazilian adults (male and female). *Journal of Thermal Biology* 2014, 42, 1–8.

38. Vardasca R, Ring EFJ, Plassmann P, Jones CD. Thermal symmetry of the upper and lower extremities in healthy subjects. *Thermology International* 2012, 22(2), 53–60.

39. Uematsu S. Symmetry of skin temperatures comparing one side of the body to the other. *Thermology* 1985, 1, 4–9.

40. Goodmann PH, Murphy MG, Siltanen GL, Kelley MP, Rucker L. Normal temperature asymmetry of the back and extremities by computer-assisted infrared imaging. *Thermology* 1988, 1, 194–202.

41. Uematsu S, Edwin DH, Jankel WR, Kozikowski J, Trattner M. Quantification of thermal asymmetry. Part 1: Normal values and reproducibility. *Journal of Neurosurgery* 1988, 69, 552–555.

42. Niu HH, Lui PW, Hu JS, Ting CK, Yin YC, Lo YL, Liu L, Lee TY. Thermal symmetry of skin temperature: normative data of normal subjects in Taiwan. *Chinese Medical Journal (Taipei)* 2001, 64, 459–468.

43. Ammer K, Engelbert B, Kern E. The determination of normal temperature values of finger joints. *Thermology International* 2002, 12, 23–33.

44. Ammer K. Temperature of the human knee: A review. *Thermology International* 2012, 22(4), 137–151.

45. International Organisation for Standardisation, ISO 9886. Ergonomics - Evaluation of thermal strain by physiological measurements. International Organization for Standardization Geneva, Switzerland, 2004.

46. Maniar N, Bach AJE, Stewart IB, Costello JT. The effect of using different regions of interest on local and mean skin temperature. *Journal of Thermal Biology* 2015, 49, 33–38.

47. Ammer K. Does the type of skin temperature distribution matter? *Thermology International* 2016, 26(Suppl), S3–S4.

48. Havenith G. Thermal conditions measurement. In: Stanton NA, Hedge A, Brookhuis K, Salas D, Hendrick HW, eds, *Handbook of Human Factors and Ergonomics Methods*, CRC Press, 2004, Chapter 60.

49. Hasgall PA, Di Gennaro F, Baumgartner C, Neufeld E, Gosselin MC, Payne D, Klingenböck A, Kuster N. IT'IS Database for thermal and electromagnetic parameters of biological tissues. Version 3.0, September 1, 2015.

50. Cannon B, Nedergaard J. Brown adipose tissue: Function and physiological significance. *Physiological Reviews* 2004, 84, 277–359.

51. Qi H, Kuruganti PT, Liu Z. Early detection of breast cancer using thermal texture maps. In: *Proceedings of the IEEE International Symposium on Biomedical Imaging, IEEE*, 2002, pp. 309–312.

52. Lin PH, Echeverria A, Poi MJ. Infrared thermography in the diagnosis and management of vasculitis. *Journal of Vascular Surgery Cases and Innovative Techniques* 2017, 3, 112–114.

53. Lee P, Ho KK, Greenfield JR. Hot fat in a cool man: Infrared thermography and brown adipose tissue. *Diabetes Obesity and Metabolism* 2011, 13, 92–93.

54. Westermann S, Stanek C, Schramel JP, Ion A, Buchner HHF. The effect of airflow on thermographically determined temperature of the distal forelimb of the horse. *Equine Veterinary Journal* 2013, 45(5), 637–641.

55. de Weerd L, Weum S, Mercer JB. The value of dynamic infrared thermography (DIRT) in perforator selection and planning of free DIEP flaps. *Annals of Plastic Surgery* 2009, 63(3), 274–279.

56. Gwinn DM, Shackelford DB, Egan DF, Mihaylova MM, Mery A, Vasquez DS, Turk BE, Shaw RJ. AMPK phosphorylation of raptor mediates a metabolic checkpoint. *Molecular Cell* 2008, 30(2), 214–226.

57. Tseng YH, Cypess AM, Kahn C R. Cellular bioenergetics as a target for obesity therapy. *Nature Reviews Drug Discovery* 2010, 9(6), 465–482.

58. Henry CJK. Basal metabolic rate studies in humans: Measurement and development of new equations. *Public Health Nutrition* 2005, 8(7a), 1133–1152.

59. Frankenfield D, Roth-Yousey L, Compher C, Evidence Analysis Working Group. Comparison of predictive equations for resting metabolic rate in healthy non- obese and obese adults: A systematic review. *Journal of the American Dietetic Association* 2005, 105(5), 775–789.

60. Byrne NM, Hills AP, Hunter GR, Weinsier RL, Schutz Y. Metabolic equivalent: one size does not fit all. *Journal of Applied Physiology* 2005, 99, 1112–1119.

61. Tappy L. Thermic effect of food and sympathetic nervous system activity in humans. *Reproduction Nutrition Development* 1996, 36(4), 391–397.

62. Westerterp KR. Diet induced thermogenesis. *Nutrition & Metabolism* 2004, 1(1), 5.

63. Smith RE, Horwitz BA. Brown fat and thermogenesis. *Physiological Reviews* 1969, 49(2) 330–425.

64. Hull D, Segall MM. Sympathetic nervous control of brown adipose tissue and heat production in the newborn rabbit. *The Journal of Physiology* 1965, 181(3), 458–467.

65. Heim T, Hull D, The blood flow and oxygen consumption of brown adipose tissue in the new-born rabbit. *The Journal of Physiology* 1966, 186. 42–55.

66. Dawkins MJ, Hull D. Brown adipose tissue and the response of new-born rabbits to cold. *The Journal of Physiology* 1964, 172(2), 216–238.

67. Inomata N, Toda M, Sato M, Ishijima A, Ono T. Pico calorimeter for detection of heat produced in an individual brown fat cell. *Applied Physics Letters* 2012, 100(15), 154104.

68. Rall JA. Sense and nonsense about the Fenn effect. *The American journal of physiology* 1982, 242,. H1–H6.

69. Kenny GP, Reardon FD, Zaleski W, Reardon ML, Haman F, Ducharme MB. Muscle temperature transients before, during, and after exercise measured using an intramuscular multisensory probe. *Journal of Applied Physiology* 2003, 94, 2350–2357.

70. World Health Organization. *International Classification of Functioning, Disability and Health: ICF*. WHO, Geneva, 2001.

71. Levine JA. Non-exercise activity thermogenesis – liberating the life-force. *Journal of Internal Medicine* 2007, 262, 273–287.

72. Gonzalez-Alonso J, Quistorff B, Krustup P, Bangsbo J, Saltin B. Heat production in human skeletal muscle at the onset of intense dynamic exercise. *Journal of Physiology* 2000, 524(2), 603–615.

73. Johnson JM, Minson CT, Kellogg DL. Cutaneous vasodilator and vasoconstrictor mechanisms in temperature regulation. *Comprehensive Physiology* 2014, 4, 33–89.

74. Barcroft H, Edholm OG. The effect of temperature on blood flow and deep temperature in the human forearm. *Journal of Physiology* 1943, 102, 5–20.

75. Taylor WF, Johnson JM, O'Leary D, Park MK. Effect of high local temperature on reflex cutaneous vasodilation. *Journal of Applied Physiology* 1984, 57(1), 191–196.

76. Kellogg DL Jr, Johnson JM, Kosiba WA. Baroreflex control of the cutaneous active vasodilator system in humans. *Circulation Research* 1990, 66, 1420–1426.

77. Hornberger J, Grimes K, Naumann M, Glaser DA, Lowe NJ, Naver H, et al. Recognition, diagnosis, and treatment of primary focal hyperhidrosis. *Journal of the American Academy of Dermatology* 2004, 51(2), 274–286.

78. Wenzel FG, Horn TD. Non-neoplastic disorders of the eccrine glands. *Journal of the American Academy of Dermatology* 1998, 38(1), 1–20.

79. Wilke K, Martin A, Terstegen L, Biel SS. A short history of sweat gland biology. *International Journal of Cosmetic Science* 2007, 29(3), 169–179.

80. Saga K. Structure and function of human sweat glands studied with histochemistry and cytochemistry. *Progress in Histochemistry and Cytochemistry* 2002, 37(4), 323–386.

81. Farrell PM, White TB, Ren CL, Hempstead SE, Accurso F, Derichs N, et al. Diagnosis of cystic fibrosis: Consensus guidelines from the Cystic Fibrosis Foundation. *The Journal of Pediatrics* 2017, 181, S4–S1.

82. Lu CP, Polak L, Rocha AS, Pasolli HA, Chen SC, Sharma N, et al. Identification of stem cell populations in sweat glands and ducts reveals roles in homeostasis and wound repair. *Cell* 2012, 150(1), 136–150.

83. Lindsay SL, Holmes S, Corbett AD, Harker M, Bovell DL. Innervation and receptor profiles of the human apocrine (epitrichial) sweat gland: Routes for intervention in bromhidrosis. *British Journal of Dermatology* 2008, 159, 653–660.

84. Harker M. Psychological sweating: A systematic review focused on aetiology and cutaneous response. *Skin Pharmacology and Physiology* 2013, 26(2), 92–100.

85. Machado-Moreira CA, Taylor NA. Thermogenic and psychogenic recruitment of human eccrine sweat glands: Variations between glabrous and non-glabrous skin surfaces. *Journal of Thermal Biology* 2017, 65, 145–152.

86. Groscurth P. Anatomy of sweat glands. In: Kreyden OP, Böni R, Burg G, eds, *Hyperhidrosis and Botulinum Toxin in Dermatology*. Karger, Basel, 2002, pp. 1–9.

87. Machado-Moreira CA, McLennan PL, Lillioja S, van Dijk W, Caldwell JN, Taylor NA. The cholinergic blockade of both thermally and non-thermally induced human eccrine sweating. *Experimental Physiology* 2012, 97(8), 930–942.

88. Bulmer MG, Forwell GD. The concentration of sodium in thermal sweat. *The Journal of Physiology* 1956, 132(1), 115–122.

89. Bovell D. The human eccrine sweat gland: Structure, function and disorders. *Journal of Local and Global Health Science* 2015, 5.

90. Smith CJ, Johnson JM. Responses to hyperthermia. Optimizing heat dissipation by convection and evaporation: Neural control of skin blood flow and sweating in humans. *Autonomic Neuroscience* 2016, 196, 25–36.

91. Taylor NA, Machado-Moreira CA. Regional variations in transepidermal water loss, eccrine sweat gland density, sweat secretion rates and electrolyte composition in resting and exercising humans. *Extreme Physiology & Medicine* 2013, 2(1), 4.

92. Baker LB, Barnes KA, Anderson ML, Passe DH, Stofan JR. Normative data for regional sweat sodium concentration and whole-body sweating rate in athletes. *Journal of Sports Sciences* 2016, 34(4), 358–368.

93. Houdas Y, Colin,Timbal J, Boutelier C, Guieu J-D. Skin temperatures in warm environments and the control of sweat evaporation. *Journal of Applied Physiology* 1972, 33(1), 99–104.

94. Shapiro Y, Pandolf KB, Goldman RF. Predicting sweat loss response to exercise, environment and clothing. *European Journal of Applied Physiology and Occupational Physiology* 1982, 48(1), 83–96.

95. Shapiro Y, Moran D, Epstein Y, Stroschein L, Pandolf KB. Validation and adjustment of the mathematical prediction model for human sweat rate responses to outdoor environmental conditions. *Ergonomics* 1995, 38(5), 981–998.

96. Gonzalez RR, Cheuvront SN, Montain SJ, Goodman DA, Blanchard LA, Berglund LG, Sawka MN. Expanded prediction equations of human sweat loss and water needs. *Journal of Applied Physiology* 2009, 107, 379 –388.

97. Havenith G, Bröde P, den Hartog E, Kuklane K, Holmer I, Rossi RM, Richards M, Farnworth B, Wang X. Evaporative cooling: Effective latent heat of evaporation in relation to evaporation distance from the skin. *Journal of Applied Physiology* 2013, 114, 778–785.

98. Chan YK, Mamat M. Management of heat stroke. *Trends in Anaesthesia and Critical Care* 2015, 5(2), 65–69.

99. Romanovsky AA, Almeida MC, Aronoff DM, Ivanov AI, Konsman JP, Steiner AA, Turek VF. Fever and hypothermia in systemic inflammation: Recent discoveries and revisions. *Frontiers in Bioscience* 2005, 10, 2193–2216.

100. Evans SS, Repasky EA, Fisher DT. Fever and the thermal regulation of immunity: The immune system feels the heat. *Nature Reviews Immunology* 2015, 15(6), 335.

101. Kurazumi Y, Tsuchikawa T, Matsubara N, Horikoshi T. Effect of posture on the heat transfer areas of the human body. *Building and Environment* 2008, 43(10), 1555–1565.

102. Scholander PF, Walters V, Hock R, Irving L. Body insulation of some arctic and tropical mammals and birds. *The Biological Bulletin* 1950, 99(2), 225–236.

103. Taylor NA, Machado-Moreira CA, van den Heuvel AM, Caldwell JN. Hands and feet: Physiological insulators, radiators and evaporators. *European Journal of Applied Physiology* 2014, 114(10), 2037–2060.

104. Krajnak K, Dong RG, Flavahan S, Welcome D, Flavahan NA. Acute vibration increases α2c-adrenergic smooth muscle constriction and

alters thermosensitivity of cutaneous arteries. *Journal of Applied Physiology* 2006, 100, 1230–1237.

105. Daanen HAM. Finger cold-induced vasodilation: A review. *European Journal of Applied Physiology* 2003, 89(5), 411–426.

106. Harris AM, MacBride LR, Foster RC, McCrady SK, Levine JA. Does non-exercise activity thermogenesis contribute to non-shivering thermogenesis? *Journal of Thermal Biology* 2006, 31(8), 634–663.

107. Sessler DI. Perioperative thermoregulation and heat balance. *The Lancet* 2016, 387(10038), 2655–2664.

108. Lowell BB, Spiegelman BM. Towards a molecular understanding of adaptive thermogenesis. *Nature* 2000, 404(6778), 652–660.

109. Morrison SF. 2010 Carl Ludwig Distinguished Lectureship of the APS Neural Control and Autonomic Regulation Section: Central neural pathways for thermoregulatory cold defense. *Journal of Applied Physiology* 2011, 110, 1137–1149.

110. Nakamura K, Morrison SF. Central efferent pathways for cold-defensive and febrile shivering. *The Journal of Physiology* 2011, 589(14), 3641–3658.

111. Silva JE. The thermogenic effect of thyroid hormone and its clinical implications. *Annals of Internal Medicine* 2003, 139(3), 205–213.

112. Silva JE. Thermogenic mechanisms and their hormonal regulation. *Physiological Reviews* 2006, 86(2), 435–464.

113. Haman F, Blondin DP, Imbeault MA, Maneshi A. Metabolic requirements of shivering humans. *Frontiers in Bioscience* 2010, 2, 1155–1168.

114. Zafren K, Giesbrecht GG, Danzl DF, Brugger H, Sagalyn EB, Walpoth B, Weiss EA, Auerbach PS, McIntosh SE, Némethy M, McDevitt M, Dow J, Schoene RB, Rodway GW, Hackett PH, Bennett BL, Grissom CK. Wilderness Medical Society practice guidelines for the out-of-hospital evaluation and treatment of accidental hypothermia: 2014 update. *Wilderness & Environmental Medicine* 2014, 25(4), S66–S85.

115. Badjatia N, Strongilis E, Gordon E, Prescutti M, Fernandez L, Fernandez A, Buitrago M, Schmidt JM, Ostapkovich ND, Mayer SA. Metabolic impact of shivering during therapeutic temperature modulation: The Bedside Shivering Assessment Scale. *Stroke* 2008, 39(12), 3242–3247.

116. Olson DM, Grissom JL, Williamson RA, Bennett SN, Bellows ST, James ML. Interrater reliability of the bedside shivering assessment scale. *American Journal of Critical Care* 2013, 22(1), 70–74.

117. Mahmood MA, Zweifler RM. Progress in shivering control. *Journal of the Neurological Sciences* 2007, 261(1), 47–54.

118. Lyman CP, Willis JS, Malan A, Wang LCH. *Hibernation and Torpor in Mammals and Birds.* Academic Press, London, 1982.

119. Brown DJ, Brugger H, Boyd J, Paal P. Accidental hypothermia. *New England Journal of Medicine* 2012, 367(20), 1930–1938.

120. Durrer B, Brugger H, Syme D. The medical on-site treatment of hypothermia: ICAR-MEDCOM recommendation. *High Altitude Medicine & Biology* 2003, 4(1), 99–103.

121. Kempainen RR, Brunette DD. The evaluation and management of accidental hypothermia. *Respiratory Care* 2004, 49(2), 192–205.

122. Doshi HH, Giudici MC. The EKG in hypothermia and hyperthermia. *Journal of Electrocardiology* 2015, 48(2), 203–209.

123. Stocks JM, Taylor NAS, Tipton MJ, Greenleaf JE. Human physiological responses to cold exposure. *Aviation, Space, and Environmental Medicine* 2004, 75, 444 –457.

124. Paal P, Gordon L, Strapazzon G, Maeder MB, Putzer G, Walpoth B, Wanscher M, Brown D, Holzer M, Broessner G, Brugger H. Accidental hypothermia–an update. *Scandinavian Journal of Trauma, Resuscitation and Emergency Medicine* 2016, 24(1), 111.

125. Walpoth BH, Walpoth-Aslan BN, Mattle HP, Radanov BP, Schroth G, Schaeffler L, Fischer AP, von Segesser L, Althaus U. Outcome of survivors of accidental deep hypothermia and circulatory arrest treated with extracorporeal blood warming. *The New England Journal of Medicine* 1997, 337(21), 1500–1505.

126. Pasquier M, Hugli O, Paal P, Darocha T, Blancher M, Husby P, Silfvast T, Carron P-N, Rousson V. Hypothermia outcome prediction after extracorporeal life support for hypothermic cardiac arrest patients: The HOPE score. *Resuscitation* 2018, 126, 58–64.

127. Tikuisis P. Predicting survival time for cold exposure. *International Journal of Biometeorology* 1995, 39, 94–102.

128. Power J, Ré AS, Barwood M, Tikuisis P, Tipton M. Reduction in predicted survival times in cold water due to wind and waves. *Applied Ergonomics* 2015, 49, 18–24.

129. Giesbrecht GG. Cold stress, near drowning and accidental hypothermia: A review. *Aviation, Space, and Environmental Medicine* 2000, 71(7), 733–752.

130. Althaus U, Aeberhard P, Schüpbach P, Nachbur BH, Mühlemann W. Management of profound accidental hypothermia with cardio respiratory arrest. *Annals of Surgery* 1982, 195(4), 492–495.

131. Tipton MJ. The initial responses to cold-water immersion in man. *Clinical Science* 1989, 77, 581–588.

132. Tipton M. Cold Water immersion. In: Tipton M, Wooler A, eds, *The Science of Beach Lifeguarding*, CRC Press, 2016, Chapter 6, pp. 87–98.

133. Cooper KE. Physiology of hypothermia. *British Journal of Anaesthesia*. 1959, 31, 96–105.

134. Rogers I. Which rewarming therapy in hypothermia? A review of the randomised trials. *Emergency Medicine* 1997, 9, 213–220.

135. Lampe JW, Becker LB. State of the art in therapeutic hypothermia. *Annual Review of Medicine* 2011, 62, 79–93.

136. Campos JM, Paniagua P. Hypothermia during cardiac surgery. *Best Practice & Research Clinical Anaesthesiology* 2008, 22(4), 695–709.

137. Callaway CW, Donnino MW, Fink EL, Geocadin RG, Golan E, Kern KB, Leary M, Meurer WJ, Peberdy MA, Thompson TM, Zimmerman JL. Part 8: post–cardiac arrest care: 2015 American Heart Association Guidelines Update for Cardiopulmonary Resuscitation and Emergency Cardiovascular Care. *Circulation* 2015, 132(Suppl 2), S465–S482.

138. Jauch EC, Saver JL, Adams HP Jr, Bruno A, Connors JJ, Demaerschalk BM, Khatri P, McMullan PW Jr, Qureshi AI, Rosenfield K, Scott PA, Summers DR, Wang DZ, Wintermark M, Yonas H. On behalf of the American Heart Association Stroke Council, Council on Cardiovascular Nursing, Council on Peripheral Vascular Disease, and Council on Clinical Cardiology. Guidelines for the early management of patients with acute ischemic stroke: A guideline for healthcare professionals from the American Heart Association/American Stroke Association. *Stroke* 2013, 44, 870–947.

139. Wyllie J, Bruinenberg J, Roehr CC, Rüdiger M, Trevisanuto D, Urlesberger B. European Resuscitation Council Guidelines for Resuscitation 2015: Section 7. Resuscitation and support of transition of babies at birth. *Resuscitation* 2015, 95, 249–263.

140. Zhang W, Ma J, Danzeng Q, Tang Y, Lu M, Kang Y. Safety of moderate hypothermia for perinatal hypoxic-ischemic encephalopathy: A meta-analysis. *Pediatric Neurology* 2017, 74, 51–61.

141. Alkabie S, Boileau AJ. The role of therapeutic hypothermia after traumatic spinal cord injury: A systematic review. *World Neurosurgery* 2016, 86, 432–449.

142. Georgiou AP, Manara AR. Role of therapeutic hypothermia in improving outcome after traumatic brain injury: A systematic review. *British Journal of Anaesthesia* 2013, 110(3), 357–367.

143. Shah M, Parikh K, Patel B, Agarwal M, Garg L, Agrawal S, Arora S, Patel N, Patel N, Frishman WH. Use of therapeutic hypothermia among patients with coagulation disorders: A nationwide analysis. *Resuscitation* 2018, 124, 35–42.

144. Taylor NA. Ethnic differences in thermoregulation: Genotypic versus phenotypic heat adaptation. *Journal of Thermal Biology* 2006, 31(1–2), 90–104.

145. Gosling SN, Bryce EK, Dixon PG, Gabriel KM, Gosling EY, Hanes JM, Hondula DM, Liang L, Bustos Mac Lean PA, Muthers S, Nascimento ST, Petralli M, Vanos JK, Wanka ER. A glossary for biometeorology. *International Journal of Biometeorology* 2014, 58(2), 277–308.

146. Greenleaf JE, Greenleaf CJ. Human acclimation and acclimatization to heat. NASA TM X-62, 008, 1970.

147. Hart JS. Climatic and temperature induced changes in the energetics of homeotherms. *Revue Canadienne de la Biologie* 1957, 16, 133–174.

148. Eagan, CJ. Introduction and terminology. *Federation Proceedings*. 1963, 22, 930–932.

149. Boeckmann M, Rohn I. Is planned adaptation to heat reducing heat-related mortality and illness? A systematic review. *BMC Public Health* 2014, 14(1), 1112.

150. Taylor NAS. Principles and Practices of Heat Adaptation. *Journal of the Human-Environmental System* 2000, 4(1), 11–22.

151. Daanen HA, Racinais S, Périard JD. Heat acclimation decay and re-induction: A systematic review and meta-analysis. *Sports Medicine* 2018, 48, 409–430.

152. Périard JD, Racinais S, Sawka MN. Adaptations and mechanisms of human heat acclimation: Applications for competitive athletes and sports. *Scandinavian Journal of Medicine & Science in Sports* 2015, 25(Suppl 1), 20–38.

153. Gibson OR, Mee JA, Taylor L, Tuttle JA, Watt PW, Maxwell NS. Isothermic and fixed-intensity heat acclimation methods elicit equal increases in Hsp72 mRNA. *Scandinavian Journal of Medicine & Science in Sports* 2015, 25(S1), 259–268.

154. Mercer JB. Cold—an underrated risk factor for health. *Environmental Research* 2003, 92(1), 8–13.

155. Mäkinen TM. Different types of cold adaptation in humans. *Frontiers in Bioscience* 2010, 2, 1047–1067.

156. Daanen HA, Van Marken Lichtenbelt WD. Human whole body cold adaptation. *Temperature* 2016, 3(1), 104–118.

157. Castellani JW, Young AJ. Human physiological responses to cold exposure: Acute responses and acclimatization to prolonged exposure. *Autonomic Neuroscience: Basic and Clinical* 2016, 196, 63–74.

158. Cheung SS, Daanen HA. Dynamic adaptation of the peripheral circulation to cold exposure. *Microcirculation* 2012, 19(1), 65–77.

Chapter 4

Infrared Thermal Imagers

The thermometric spectrum that was described by William Herschel in his 1800 experiment is today termed the infrared radiation. This is the definable part of the electromagnetic spectrum which is beyond the red end of the visible light. As described in Chapter 1, a long period elapsed before sensitive sensors had been developed to make thermal imaging a usable tool.

The infrared spectrum has been subdivided into three main sections, short or near, 750 nm–2.5 μm, medium or mid 2.5–10 μm, and long wave or far 10–300 μm (Fig. 4.1). In medicine and biology 3–8 μm and 8–15 μm wavelengths are used for temperature measurement.

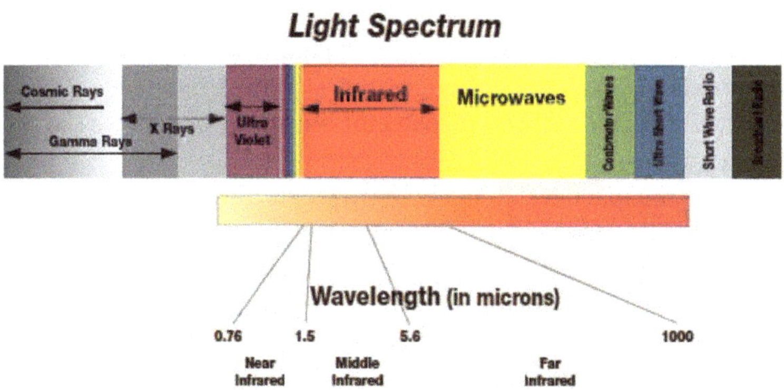

Figure 4.1 Position of infrared radiation in the electromagnetic spectrum.

The Thermal Human Body: A Practical Guide to Thermal Imaging
Kurt Ammer and Francis Ring

ISBN 978-981-4745-82-6 (Hardcover), 978-0-429-01998-2 (eBook)
www.jennystanford.com

4.1 Detectors

There are two main types of detectors used for thermal imaging: *photon detectors* and *thermal detectors* [1].

In *photon detectors* the radiation is absorbed within the material by interaction with electrons. The electrical output signal results from the changed electronic energy distribution. Photon detectors show a selective wavelength dependence of the response per unit incident radiation power. They give both perfect signal-to-noise performance and a very fast response. This, however, is achieved only by cryogenic cooling. Photon detectors can be further subdivided into different types depending on the type of interaction. The most important are: intrinsic detectors, extrinsic detectors, photoemissive (metal silicide Schottky barriers) detectors, and quantum well detectors.

The second class of IR detectors consists of *thermal detectors.* In a thermal detector the incident radiation is absorbed to change the temperature of the material. The resultant changes in some of the physical properties are used to generate an electrical output. The detector element itself is suspended on lags, which are connected to the heat sink. The thermal effects are generally wavelength independent; the signal depends upon the radiant power (or its rate of change) but not upon its spectral content. Pyroelectric detectors measure a change in the internal spontaneous polarization, but bolometers measure a change in the electrical resistance. Unlike photon detectors, thermal detectors usually operate at room temperature.

The principle developments in such detectors began during the 1940s driven by the needs of the military. However, it was not until 1958 that the military exclusion for infrared imaging was lifted, and medicine along with many other industrial applications was free to develop. Some of the changes in technology over the last half century have had a bearing on the quality and use of thermal imaging in medicine. In consulting earlier papers and books on this subject, this must be taken into account.

Infrared sensor technology is an aspect of thermal imaging that has dramatically improved over the last 50 years. The early systems for thermal imaging were based around a single element detector, usually indium antimonide (2–5 microns). The detector was mounted at the base of a Dewar vessel, which was filled with liquid nitrogen. Using a single detector in this way required an optical mechanical scanning system to build up an image (Fig. 4.2).

The use of liquid nitrogen to cool the detector had limitations. The camera system was unable to be angled much from horizontal. In medical applications, unless the patient was standing before the camera, some limitations were imposed by camera angle restriction. Surface silver mirrors were introduced, allowing the patient to be imaged from an examination couch. This also had its limitations especially if the mirror was influence by reflected radiation. Eventually more sustainable cooling systems using thermoelectric (Peltier) or gas expansion systems such as Stirling cooling were employed. This then meant that many of the positioning limitations were overcome.

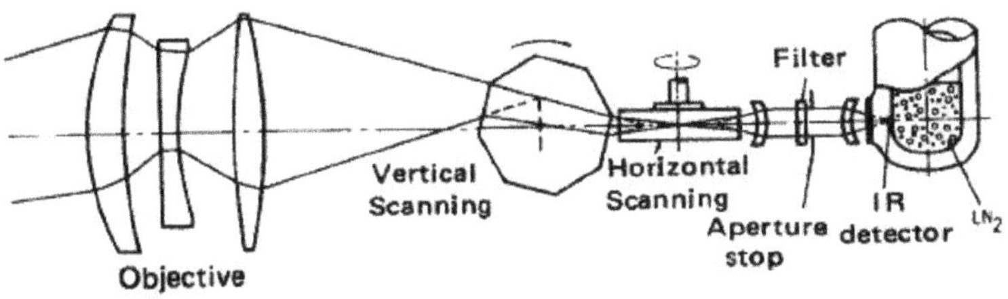

Figure 4.2 The optical mechanical scanner with reflecting optics of the first imagers.

In the single detector system, the image is built up from the two directional scans (Fig. 4.3).

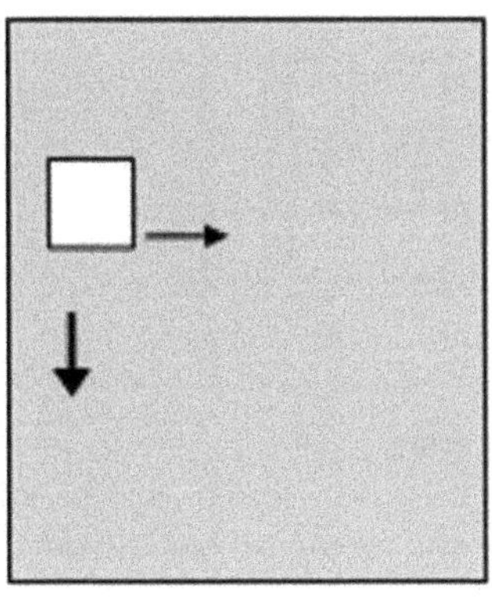

Figure 4.3 Single-detector scan.

The practical outcome is that speed and resolution are dependant, so the higher the scanning speed, the lower the resolution of the final image. Attempts to improve this were made by a multi-element detector. This resulted in shorter scan times and improved resolution; however, the problem of alignment of the elements meant that resolution was ultimately compromised (Fig. 4.4).

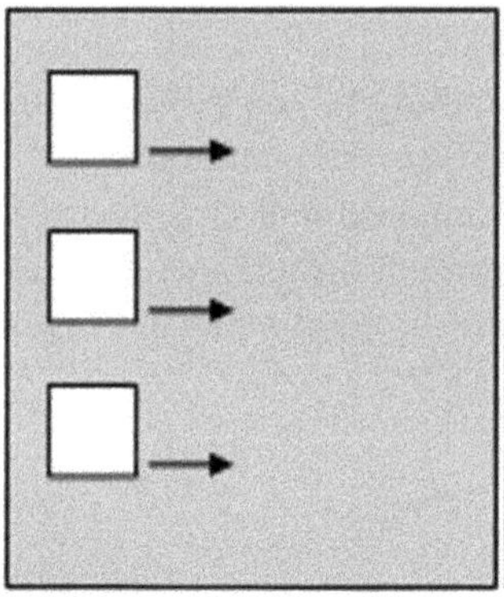

Figure 4.4 Multi-element scan.

Finally, focal plane array technology provided a large matrix of detector elements that could be scanned electronically, resulting in both high speed and high resolution (Fig. 4.5).

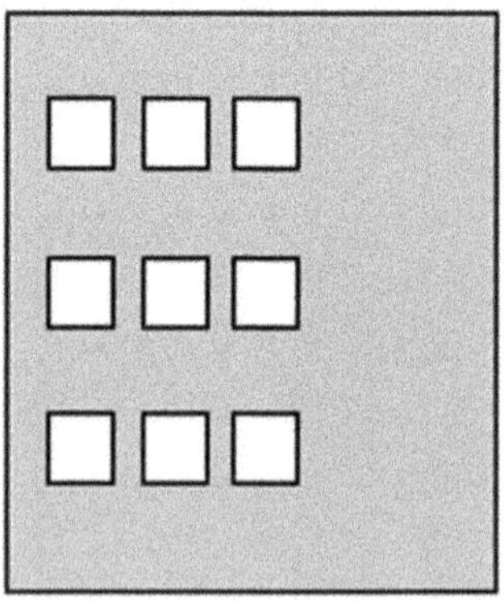

Figure 4.5 Focal plane array.

The most successful detector has proved to be the mercury cadmium telluride (HgCdTe) with its wide range and fast response.

Its performance is most reliable at the short and medium IR wavelengths that are relevant to medical thermography.

The first single-element detectors (Fig. 4.3) were followed by multi-element systems (Fig. 4.4). These in turn were replaced by focal plane array systems (Fig. 4.5) providing real-time images.

There are now a wide range of electronic sensors, and thermal cameras built around uncooled bolometers can be effective for medical use. Imaging with uncooled detectors is now well established, and microbolometer arrays are the most used technology. Present state-of-the-art microbolometers are based on polycrystalline or amorphous materials, usually vanadium oxide (Vox) or amorphous silicon (a-Si).

Table 4.1 shows a list of some of the sensors that are used for infrared detection.

Table 4.1 Sensors used for infrared detection

Detector	Spectral range (μm)
Indium gallium arsenide (InGaAs) photodiodes	0.7–2.6
Germanium photodiodes	0.8–1.7
Lead sulphide (PbS) photoconductive	1.0–3.2
Lead selenide (PbSe) photoconductive	1.5–5.2
Indium arsenide (InAs) photovoltaic	1.0–3.8
Platinum silicide (PtSi) photovoltaic	1.0–5.0
Indium antimonide (InSb) photoconductive	2.0–5.0
Mercury cadmium telluride (MCT, HgCdTe) photoconductive	0.8–25.0
Mercury zinc telluride (MZT, HgZnTe) photoconductive	2.5–20.0
Amorphous silicon (a-Si) (microbolometer)	8.0–13.0

4.1.1 Focal Plane Array Detectors

Multi-element electronically scanned focal plane arrays have generally replaced the smaller number of elements. This also has meant that the mechanical scanning previously used is no longer required. Such systems are also called staring arrays. It is clear that the size of each detector is important to the spatial resolution of the image, as is the number of pixels within the array. They may be currently described as low resolution 160 × 120 (or less) pixels, medium resolution 320 × 240, or high resolution 640 × 480 and above (Fig. 4.6).

Figure 4.6 Three examples of focal plane array IR imagers.

As lack of uniformity across the array can become a problem, and individual pixel response can vary. Pixel correction is then applied before the final image is displayed. Calibration of these systems on a regular basis is required, and regular servicing by the manufacturer will ensure that the whole array is optimized to uniformity. The user should therefore be aware that however excellent these systems are in practical daily use, their full maintenance is of great importance to overcome the image deterioration over time, which may not necessarily be evident to the user.

4.2 Lenses Used with Infrared Cameras

The ability to focus a lens system in the same way as with a photographic camera has also simplified the camera system, leading to reduction in size and simpler use (Fig. 4.7) [2]. There are a limited number of materials that can be used to transmit infrared radiation. Few are, however, stable enough for practical use. Germanium is one that is suitable especially for long-wave IR, but is expensive. Gallium arsenide and silicon, zinc sulphide (ZnS) and zinc selenide (ZnSe) are used as less expensive materials and are suitable for the medium-wave IR. To determine the materials that can be optically polished with good performance is a subject of continuing research. Where a lens is used as an objective and exposed to the user, care must be taken to keep it covered by the lens cap when not in use, and avoid any physical contact with the lens surface.

Figure 4.7 An infrared lens. *Specifications*: germanium and GASIR®; focal length 12.8 mm; aperture-based f-number f/1.00; maximum image circle 8.2 mm; waveband 8–12 μm; transmission > 97% average over waveband; focus range 0.4 m to infinity with 0.39 mm refocus (Infrared Optics Olen Belgium).

4.3 Image Acquisition

In addition to the above, there are three conditions that have an important impact on image acquisition.

4.3.1 Emissivity

As radiative temperature measurements are based on the emissivity of grey radiator, the choice of the correct emissivity of a radiating object is critical for accurate temperature readings. Most cameras provide facilities for adjusting emissivity according to the type of material being imaged. In medical applications, this must be set to 0.98, which is the emissivity of the human skin.

4.3.2 Temperature Range and Level

Many cameras work in a temperature range that can exceed 600°C. As the human is maintained within a narrow temperature range less than 25°C, lower temperature ranges such as 5, 10 or 15°C are usually adequate. The operator should be aware of the danger of automatic ranging provided by many manufacturers. The reason is that all objects in the field of view are used to find an arbitrary level. A fixed temperature range, as described above should always be used at the level of the median value of the temperature window.

4.3.3 Focus

Since the thermal camera uses an optical system, as in photography focus is most important. This is one fault that if made during image capture cannot be software corrected. If infants are being examined by thermography, their movement can make this difficult. It is in these situations that the modern high-speed imaging proves to be a great advantage. It is important that parallax distortion is avoided caused by angled views.

4.4 The Image Display

Earlier scanning systems could only produce a monochrome display. As a thermogram represents the distribution of the intensive quantity temperature, the primary image formed by the camera is always a monochrome grey-scale.

The camera systems from AGA and Bofors, both developed in Sweden in the 1960s used a monochrome oscilloscope display. To identify temperatures of selected features an electronically generated *isotherm* was provided. An isotherm is a set of points within the same range of temperature; an isothermogram is defined as the graphic recording of one or more isotherms of the thermal image [3]. Therefore, isotherms are representations of grey-scale differences which represent differences in temperature. Display technology of that time was able to differentiate grey shades.

The first false colour thermal images were created by multiple exposure photography of the isotherms ranging through the image at set intervals of temperature. Each exposure on a single camera frame passed through a different colour filter. The final result was a colour isotherm image that corresponded to the scale displayed at the edge of the thermogram.

When the short- to mid-range cadmium mercury telluride (0.8–25 microns) detector was introduced, at the same time colour monitors were introduced to directly display the image. The dependence on photographic film processing before the colour image was visible to the operator was overtaken by the immediate display on a colour monitor.

By the mid 1970s, computer processing started with a continuing series of improvements to the practical usage of the technology for infrared imaging. The images being analogue required conversion to digital. A succession of improvements in image quality, valuable opportunities opened up for quantification of temperature distribution and transformed thermography into a measurement device.

4.5 The Thermal Image

4.5.1 Definitions

Several definitions are available.

Infrared Thermography is the recording of the temperature distribution of a body from the infra-red radiation emitted by the surface [3]. However, recording temperature distribution is not the same process as measuring temperature.

Thermal Image is the visible product of thermography [4].

Thermogram is defined as a graphic recording of the thermal image obtained by thermography. This term does not describe a magnetic recording or any other storage system which is not graphic and seen by the eye. Prefixes infrared, microwave, etc., also apply to this term.

The equipment manufacturers have introduced their own trade names, which are sometimes used by authors. *Thermovision,* one of the oldest system names introduced by AGA (and AGEMA Infra Red Systems, now FLIR) can be found throughout the literature spanning some 50 years [5].

A modern reference work in the English language is The Cambridge Dictionary of Science and Technology [6]. The definition of *Thermal Imaging* is: Imaging based on the detection of weak infrared radiation from objects. Applications include the mapping of the Earths' surface from the air, weather mapping and medical thermography.

The term *thermograph* was defined in the 1977 report as the instrument used for thermography [4]. However, this word has appeared many times in publications to represent the image, not the imaging system. The Cambridge Reference describes thermograph as a continuously recording thermometer, and then describes the Wheatstone bridge resistance thermometer and the recording drum usually used for environmental monitoring.

The oldest term associated with this subject is *thermogram.* This was given by Sir John Herschel in 1840, when he made the first thermal image by solar radiation and evaporography, reported to The Royal Society London. Perhaps an important question at the present time is, should *thermal imaging* replace *thermography*, but retain *thermogram* as an image, which may be viewed electronically on a visual display unit, or as hardcopy from a photographic or electronic printing process [5]?

4.5.2 Image Content

Each manufacturer of IR cameras has its own form of image display, often including the company logo. However, there are some essentials that should always be visible, and be part of the image from time of image capture. The camera variables are usually temperature range and level. An image of the same subject recorded at a 10°C thermal range will appear quite different at a 20°C. There is usually a scale indicated at one edge of the image to indicate this.

Some systems also display an image number and certain identification parameters from the particular camera itself. In clinical thermography the correct patient identification, and date and time of recording are essential. These may often be provided by the image capture software. Thermograms are sometimes shown in publications or patient files without the critical camera settings displayed. This is bad practice, and good software will require these data before they can be digitally stored. In the old example of a thermogram shown in Fig. 4.8, the margin scale indicates that the camera was set at 5°C range and the seven coloured isotherms plus white are spread over this range. However, no absolute temperatures were indicated at this time unless a reference source was shown.

In a modern thermogram, the temperatures are displayed with the image (Fig. 4.9).

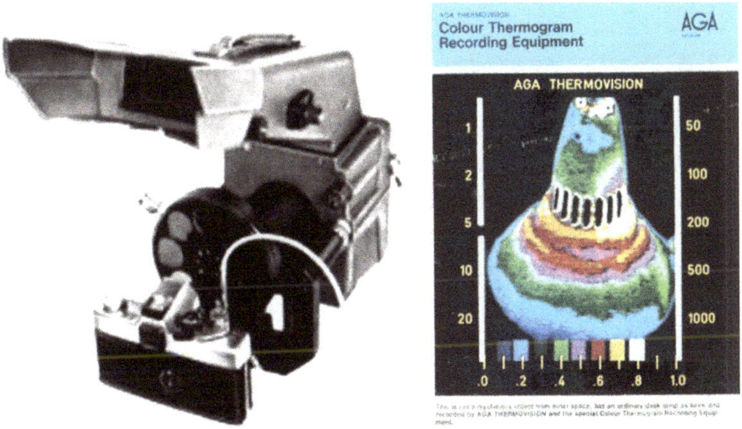

Figure 4.8 Oscilloscope camera used to create false colour thermograms and a colour thermogram of a desk lampshade made by this process.

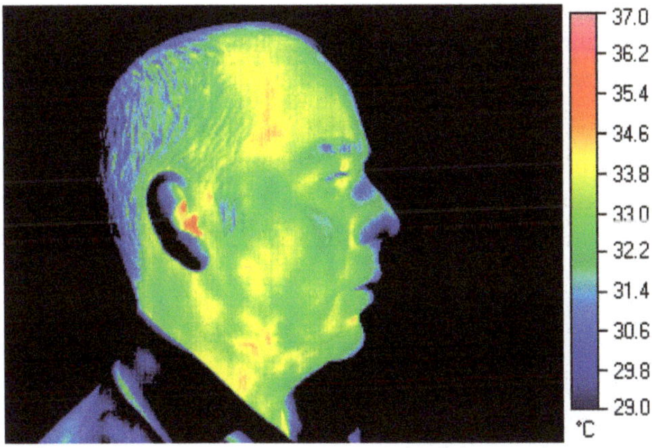

Figure 4.9 A thermal image showing the temperature scale at the right margin. This indicates that an 8°C scale had been used, the lowest temperature being 29°C and the highest 37°C.

It will be shown in a later chapter that the position of the subject is important and should be defined within a normal working protocol. So also care should be taken over the composition of the image, similar to that used in conventional photography. Note that in

Fig. 4.9, the subject occupies the larger proportion of the field of view (75–85%) and that there are no background objects at a displayed temperature to add confusion to the image.

4.5.3 Image Analysis

A number of commercial cameras are equipped with different software for processing the captured image. But not all of them meet the minimal requirements for medical thermography, because they have been designed for general purpose engineering. Minimum requirements for application in medicine and biology are as follows:

- Editable colour and monochrome palettes. This implies the ability to edit colour sequence and define temperature range.
- A minimum of four freely defined isotherms.
- Multiple spot temperatures and multidirectional thermal profiles.
- Defining multiple regions of interest with various geometrical and irregular shapes.
- Temperature measurements to be extracted are mean temperature and standard deviation, maximum and minimum temperature are optional.

The selection of regions of interest from the thermogram is an important step. Currently the automatic selection of regions of interest is not generally recommended, because image processing can easily identify the strong contrasts, e.g. the outline or the hottest or coolest spot. These strong contrasts are not always related to anatomical landmarks.

It has been shown by Ammer that unless a code of rules such as anatomical limits for each region are followed, the measurements extracted from that region cannot be repeated even by the same operator [7]. This is a discipline that can be incorporated in image analysis software.

If care is taken to exactly register a series of images, then expressions of change over time can be made. This can be achieved again by software using a template designed around the first image, and each subsequent image must be aligned to the same template, usually by careful repositioning of the patient, and adjustment of the camera to subject distance to fit the template. This technique can be recommended for sequential follow up visits by the same patient, and makes each image more comparable, and with the same pixel count in a region of interest. From the above it will be seen that consistent working with the camera and the software are the key to good technique and necessary to obtain reproducible results.

References

1. Rogalski A. Infrared detectors: an overview. *Infrared Physics & Technology* 2002, 43, 187–210.
2. Grenn MW, Vizgaitis J, Pellegrino JG, Percont P. Infrared camera and optics for medical applications. In: Diakides M, Bronzino JD, Peterson DR, eds, *Medical Infrared Imaging: Principles and Practices*. CRC Press, Boca Raton, 2013, pp. 5.1–5.15.
3. IUPS Thermal Commission. Glossary of terms for thermal physiology. *The Japanese Journal of Physiology* 2001, 51(2), 245–280.
4. Aarts NJM, et al. Thermographic Terminology. *Acta Thermographica* 1978, Supplement 2.
5. Ring EFJ. Terminology. Is Thermography an Accurate Word? *European Journal of Thermology* 1998, 8(1), 5–6.
6. Walker PMB. *Cambridge Dictionary of Science & Technology*. Cambridge University Press, Cambridge, New York & Melbourne, 1990.
7. Ammer K, Ring F. Standard procedures for infrared imaging in medicine. In: Diakides M, Bronzino JB, Peterson DR, eds, *Medical Infrared Imaging, Principles and Practices*, CRC Press, 2013, pp. 32.1–32.14.

Chapter 5

Quality Assurance Procedures and Infrared Equipment Operation

5.1 Quality Assurance Procedures

Thermal imaging is a remote sensing procedure for recording infrared radiation from an object. It is therefore necessary to carry out all available checks on the imaging system and the applied technique in order to reduce variations as far as possible. The following procedures are required to monitor system performance.

5.1.1 Calibration and External Reference Source

In the majority of clinical applications, it is useful to know how to retrieve temperature data from the image. The image itself expresses the temperature distribution of the target in two dimensions. A blackbody reference is used by the manufacturer in order to correctly set up the camera system as part of the manufacturers' product control (Figs. 5.1 and 5.2). However, in clinical practice it is a valuable addition to a thermography service to have a constant temperature source as additional equipment. Many camera systems are expected to drift over time from the factory settings, and without the use of a reference source, the operator may be unaware of such changes in performance. An advantage of thermal imaging in medicine is that patients may be monitored over time to document any changes in

The Thermal Human Body: A Practical Guide to Thermal Imaging
Kurt Ammer and Francis Ring

ISBN 978-981-4745-82-6 (Hardcover), 978-0-429-01998-2 (eBook)
www.jennystanford.com

a clinical condition, or in response to a course of therapy. In such cases, it is even more important to be able to be assured that changes found on the thermal image are indicating clinical or physiological change, not camera drift.

A number of commercial reference sources are available and many are sold by the camera manufacturers themselves. They may be cylindrical in outer shape but will have an opening which will provide the calibration target area. The inner surface coating will be of high emissivity to achieve the highest value.

If the reference source can be mounted close to the patient and in the same focal plane, a constant source at a fixed temperature (such as 30°C) can be incorporated in the image. This is especially useful when an uncooled bolometric imager is being used. Some larger reference sources are not designed for such use, but can be trolley mounted, and therefore moved into the field of view when required [1, 2, 7–9].

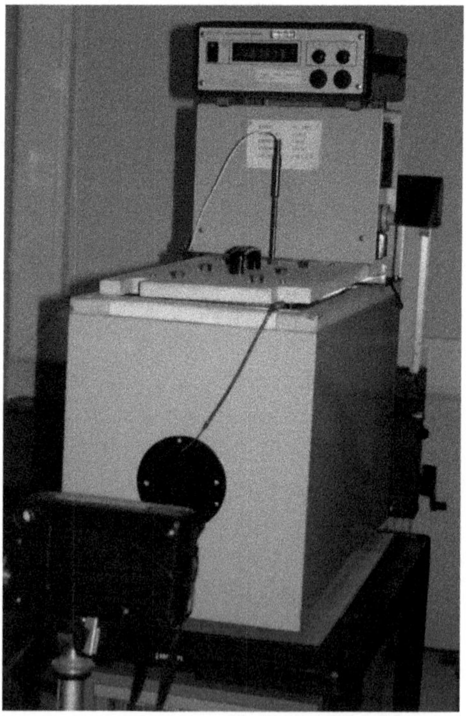

Figure 5.1 A calibration laboratory blackbody, National Physical Lab, UK.

Figure 5.2 A black body reference source suitable for regular use.

5.1.2 Time to Reach Camera Stability

An important factor in thermal imaging is that many camera systems have a period after switching on the power, when they are not fully stabilized. In some cases, the manufacturer indicates this by not displaying an image at this stage. Even then, a given camera may require a longer time to reach its optimum stability. This can be determined by recording a constant temperature source over time, and expressing the data as a graph. This should be performed at intervals of 3–6 months, since some cameras will change in their performance over time.

It is always necessary to control the ambient temperature of the room in which thermographic examinations are performed. When the start up to stability test is being carried out, the room temperature should also be known, be stable and recorded with the measurement data record. Keeping these records on a regular basis will be useful in detecting any tendency to change that will require instrument maintenance.

5.1.3 Offset Drift

When a new IR camera is purchased, it will have been calibrated by the manufacturer to their reference source. It may be that when it has been installed with a locally purchased reference source that a small difference in temperature measurement is found. This should be documented for future reference. The particular offset may remain constant for over a year or it may increase. A useful test is to set up the camera from the start before the blackbody reference, and regularly record images and measurements from the region of interest placed over the thermal image of the heated reference. Different cameras may show different response over the duration of the test. Figure 5.3 shows a graph of four different thermal cameras tested as described against the same reference [2].

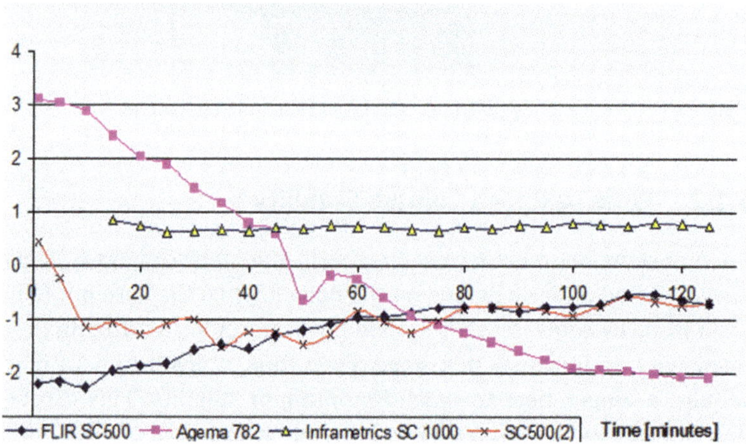

Figure 5.3 A plot of the offset drift of four IR cameras, the Agema 782 was a liquid nitrogen cooled camera, probably with a failing cooling system. The zero on the vertical axis is 30°C, at which the reference was set. The horizontal axis is time in minutes from the power on. It will be seen that one camera was 1°C below the reference at 60 min.

This test provides further evidence that IR cameras should be regularly maintained, because the offset drift for a particular instrument cannot be guaranteed to remain constant over time.

Some cameras need a shutter mechanism to remain calibrated, and drift can occur for a few seconds at the time of the shutter action. This should normally be very brief and have a minimal effect. In Fig. 5.4, this fluctuation in a camera in need of service is shown.

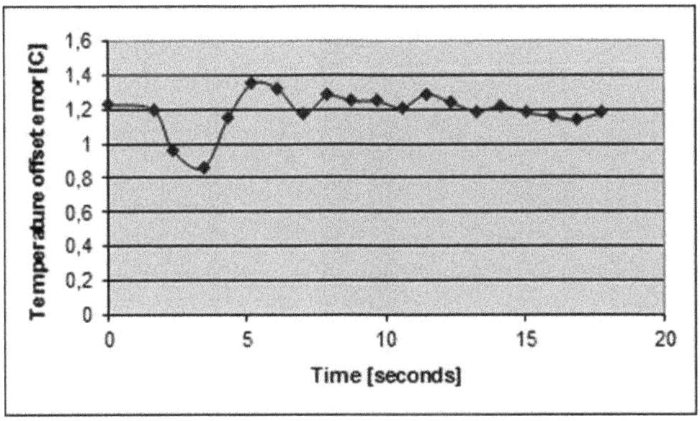

Figure 5.4 The change in offset drift in a shuttered camera during the shutter action.

5.1.4 Uniformity of the Image

All images created by use of electronic sensors are expected to be even in terms of temperature sensitivity and of focus. A focal stop is often part of the camera optical system that may exclude signals from extreme edges or corners. Nevertheless, it can become evident by regular assurance tests that, the uniformity of the image may also change over time.

A simple test can be performed with the reference source, which can be positioned at different parts of the image. A minimum of five locations should be selected, i.e. one at each corner of the image and one placed centrally. Then from the recorded image a region of interest is defined which should occupy some 70–80% of the blackbody source and the temperatures recorded. Ideally all five results should be within close tolerance <0.2 or 0.3°C according to the camera specification. These results will also be distance dependent, as the greater the distance from source to camera, the smaller the registered image. Depending on the size of the reference source aperture, an optimal distance should be set, say 0.5 m or 1.0 m and this should always be used in repeat tests. Figure 5.5 illustrates the value of performing this test.

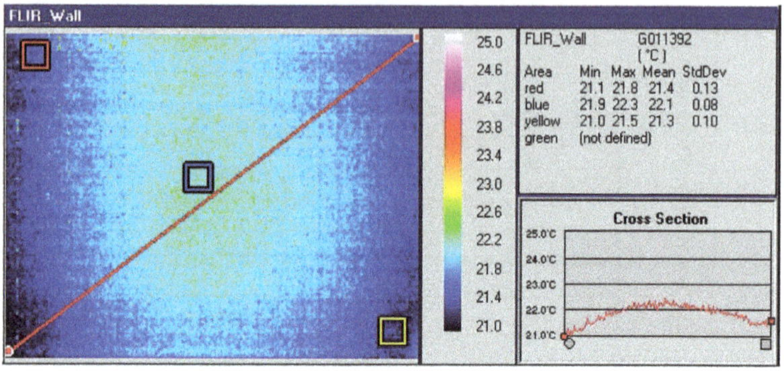

Figure 5.5 Thermogram recorded by imaging a large area of a wall within an insulated room known to be of even temperature. The failing detector is showing maximum heat in the centre of the image by 0.6°C. After repair and realignment of the detector by the manufacturer, this fault was removed.

5.1.5 Reflected Heat Test

A simple demonstration of how the camera reacts to reflected heat can also be shown in this test. When a vessel containing hot water approx 37°C is placed at the edge of the field of view the temperature measurement of the reference source may be distorted, giving a false high or low result. This demonstrates that in carrying out this test all warm objects, such as computers, lighting or personnel should be well clear of the imaging plane in order to avoid reflected heat.

Figure 5.6 illustrates the reflection error caused by flooding the detector from a hot nearby source of heat. Uncooled bolometers are most prone to this error.

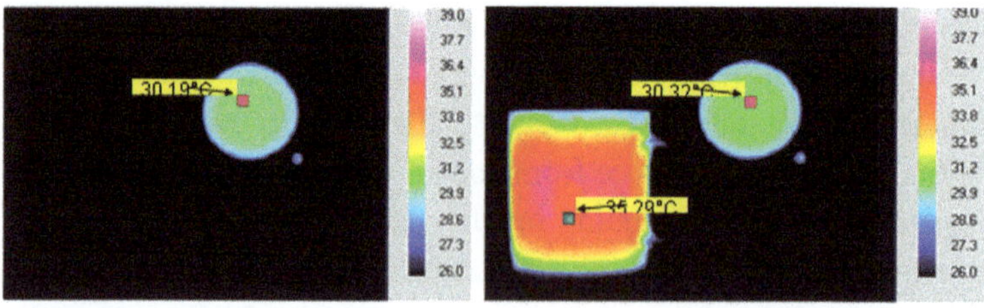

Figure 5.6 The image of a reference source recording 31.09°C on the left and after placing a warm object, the reference shows an increased temperature of 31.32°C.

In conclusion, the tests described are of important to the operator of thermal imaging procedures. The imagers have improved in performance and reliability in recent years. However, they are still more deficient in some areas than is always recognized.

In specification, there can be an offset of ± 2°C even after the manufacturer's calibration.

In stability, short-term drift is possible, and medium-term, even after power-on time specified by the manufacturer. Long-term drift can be found over several hours or days.

Uniformity of the image can become an issue over time.

Scene flooding by the presence of another heat source near the field of view or in the image can affect ± 2°C.

5.1.6 Temperature and Spatial Resolution

Resolution refers to the ability of the system to detect or "resolve" a small difference in the scene. This can be thought of as a measure of the "sensitivity" of the imaging system. The resolution of a thermal imaging camera can be spatial and thermal. In the former, it is possible to image standard charts with lines of different length and thickness. In the latter, heat sources of different size can be imaged to ascertain the smallest areas that will be registered by the camera. In both cases, the same rules that apply to photographic cameras and all other optical devices will apply. The most obvious is that distance between the target and the camera will determine the size of the target within the image and also its sharpness at optimal focus. The parameters are directly affected by the lens used, whether a narrow or wide angle type.

Spatial resolution refers to the ability of the system to spatially discriminate an object in the field of view. Since this is clearly affected by the optics in front of the detector, the spatial resolution of the detector itself is sometimes (incorrectly) specified by the size of the pixel array, since in general, more pixels improve the spatial resolution of the system. However, other factors (fill factor of the array, diffraction limits, etc.) can limit or at least play a role in the spatial resolution of the detector.

An IR camera can be tested by using a heated target plate on which a pattern of bars of varying thickness, e.g. 1–10 mm. By varying the camera to target distance and recording images, some indication will be given of how much spatial resolution can be reduced as the target distance increases. Examples can be found in [4, 5], where different cameras have been evaluated (Fig. 5.7).

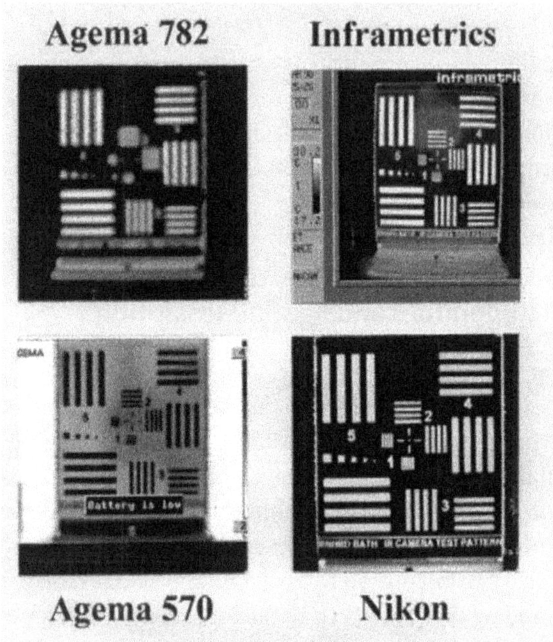

Figure 5.7 Thermograms of a heated resolution chart recorded at the same distance. The centre spot is 1 mm diameter and the bars are graded up to 5 mm thickness.

Thermal resolution is often described in terms of the NETD (noise equivalent temperature difference), or the MRTD (minimum resolvable temperature difference). Both of these criteria refer to the ability of the system to detect a small change in temperature of an object in the field of view. A simple way of testing for thermal resolution is by imaging the surface of a container of heated water to a known temperature. A mask of rigid material is placed near to the water surface, with varying size apertures. From the thermal image, the aperture size that yields temperature readings close to the known water temperature will indicate the approximate thermal resolution of the system at the distance set, and with the optical settings used.

5.2 Infrared Equipment Operation

The procedures described in this section are an important part of image acquisition.

5.2.1 Camera Position

Infrared cameras are commonly supplied and used with a photographic tripod. For the purposes of standardization, this is rarely satisfactory. The camera should be positioned before the target or subject at or close to 90° to avoid distortion errors. In the case of the human body, where almost all skin surfaces have some degree of curvature, there can only be a compromise. In some cases, multiple images at different angles from the subject may be required, e.g. the female breast. In clinical practice, many different views of the patient can be acquired with the patient standing or seated before the camera. Modern uncooled cameras can also be used in a vertical alignment when the patient is lying on an examination couch or with hands on a table (Fig. 5.8).

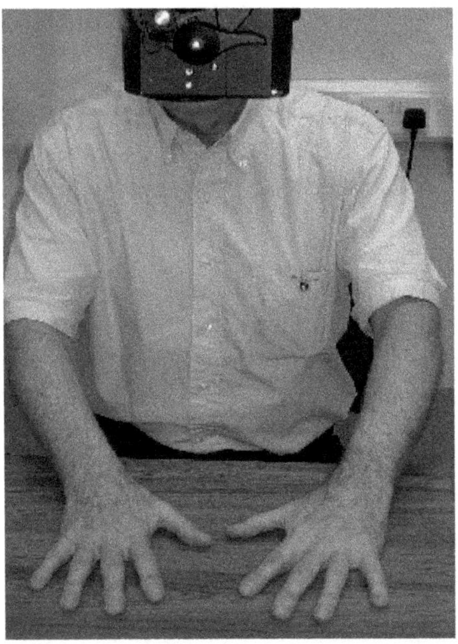

Figure 5.8 Vertical camera position.

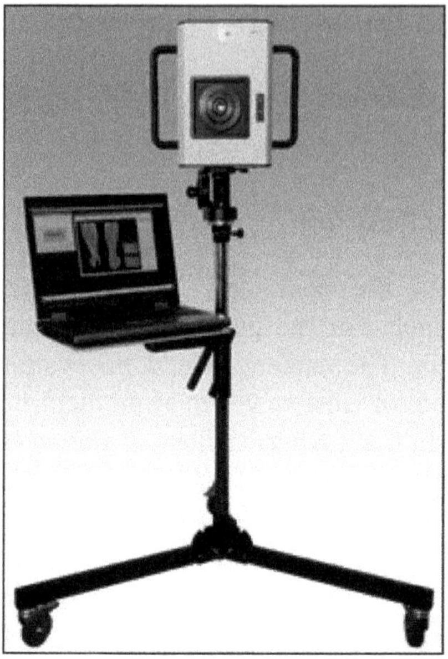

Figure 5.9 A single pillar stand for IR camera mount.

An important aid to good positioning is also in the type of camera stand that is used. In all these situations, a single pillar camera stand is ideal. Complications caused by unrepeatable angles created by the use of a ball and socket camera mount can and should be avoided. Photographic tripods often sold with infrared systems are very inconvenient for medical thermography. Apart from the time taken to change height, they are often equipped with a ball and socket head. This makes repositioning on further examinations of the subject extremely unreliable. In regular medical thermography a single pillar stand offers the best facilities, horizontal and vertical adjustments are easy, as the camera is usually counterbalanced (Fig. 5.9.). Some more expensive systems offer power assistance, but this is not required for the very modern smaller cameras.

Some examples of bad image capture are shown in Fig. 5.10. In the left image, the angle of the hand is wrong and the background has not been screened off. The centre image shows the wrong position of the knees, tight trousers (not removed) and half the image taken up with irrelevant material. Therefore, the target knees are too far away, and at the wrong angle to the camera. The right image of a face is correctly positioned.

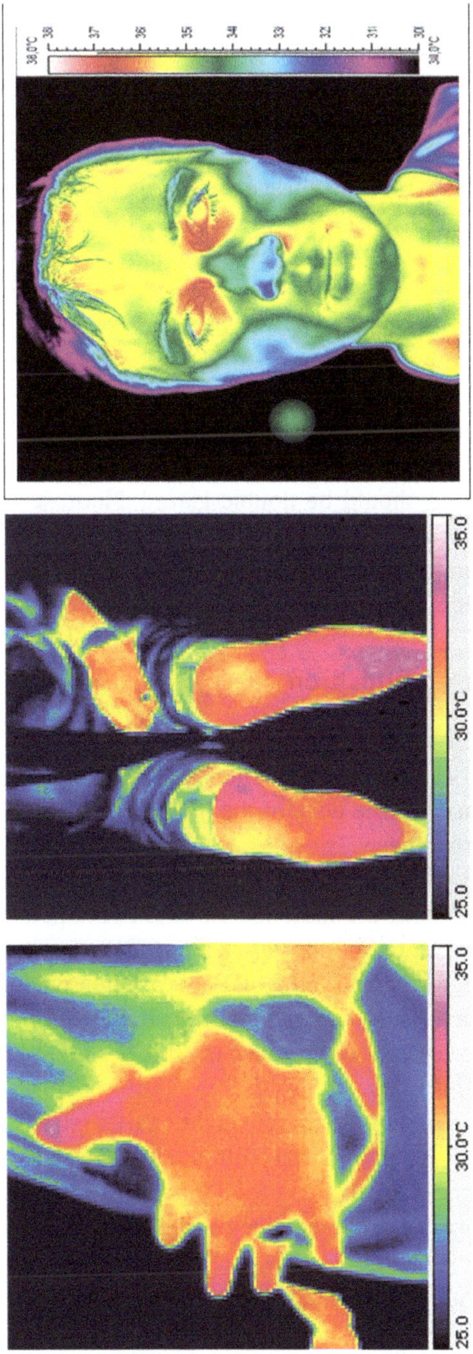

Figure 5.10 (Left) Wrong position: hand, (middle) wrong position: knees; (right) correct position: face.

Whenever possible, thermographic examination of human subjects should be performed in a controlled environment. Air conditioning that can be set to a constant temperature, typically 20°C or 22°C should be used, and may need extra baffle plates to deflect air currents away from the patients and the camera system. It may be necessary to operate this through a time switch so that the room temperature has reached stability well before (>30 min) the planned work with the camera. Room temperature must not vary more than ±0.5°C for 1 h. Furthermore, the thermal reference source described above will require a specific time period for its own stability and may be powered up for 1 h or more before use. Some reference sources can be left on continuously if required for use on a regular basis.

There are some conditions that a good thermally stable room can have. Minimal window area which may be partially blocked from sunlight and should be double glazed. Fluorescent or other cold lighting is preferable to tungsten which generates heat. A large indicating thermometer display is recommended, so that immediate indication of room temperature is highly visible.

5.2.2 Image Composition

The target area of each different thermal image should be so positioned that the ROI used for measurement should be central rather than placed at the edge of the image, in order to obtain the most accurate readings.

Also of major importance is the background, that part of the image which is outside the anatomical area to be studied. Ideally, this should be at a cooler temperature, and if possible out of temperature range, and therefore displayed a black area. Often when imaging a part of the human body, this can be difficult, but can be overcome by the use of a screen or cloth which is used as a thermal shield for those background areas.

In practice, a selection of lightweight boards (size: 0.3–0.5 m^2) can be used. Also, small material (cotton) sheets such as those used in medical surgery can also be useful for this purpose. Boards can be held underarm or between knees or feet by the patient in some cases.

For temperature level and temperature window, see Section 4.3.2.

5.2.3 Colour Scale

The primary thermal image is monochrome, with white as hot and black as cold. False colour is frequently used which helps the human eye to discriminate between close density shades of grey. It is important to note that false colour can distort the anatomical image [6]. Most thermal cameras designed for industrial use may show a colour scale described as iron. Here the hottest temperatures are white to yellow, with cooler temperatures shown in amber and red. The most useful false colour scale, however, is termed rainbow, where hot is red and cold is blue with shades in between in spectral order. Modern software systems will provide a large number of grades of each colour, 256 or 512 are possible, but maintaining the natural spectral order as observed in the rainbow across the sky. The image should always include the colour bar to indicate the false colour range used, with relevant temperatures against that scale.

References

1. Miklavec A, Pusnik I, Batagelj V, Drnovsek J. A large aperture blackbody bath for calibration of thermal imagers. *Measurement Science and Technology* 2013, 24, 025001.
2. Simpson RC, McEvoy HC, Machin G, Howell K, Naeem M, Plassmann P, Ring F, Campbell P, Song C, Tavener J, Ridley I. In-field-of-view thermal image calibration system for medical thermography applications. *International Journal of Thermophysics* 2008, 29, 1123–1130.
3. Plassmann P, Ring EFJ, Jones CD. Quality assurance of thermal imaging systems in medicine. *Thermology International* 2006, 16, 10–15.
4. Ring EFJ, Dicks JM. Spatial resolution of new thermal imaging systems. *Thermology International* 1999, 9(1), 7–14.
5. Richards RE, Allen J, Howell KJ, Smith RE. Evaluation of three thermal Imaging cameras for skin temperature measurement using a blackbody reference source and a spatial resolution test object. *Thermology International* 2013, 23(1), 17–23.
6. Ammer K. The influence of colour scale on the accuracy of infrared thermal images. *Thermology International* 2014, 24(4), 141–146.
7. Machin G, Chu B. High quality blackbody sources for infra-red thermometry and thermography between –40°C and 1000°C. *Imaging Science* 2000, 48, 15–22.

8. Simpson R, Machin G, McEvoy HC, Rusby RL. Traceability and calibration in temperature measurement: A clinical necessity. *Journal of Medical Engineering & Technology* 2006, 30, 212–217.

9. Machin G, Simpson RC, Broussely M. Calibration and validation of thermal imagers. *QIRT Journal* 2009, 6, 133–147.

Chapter 6

Standard Protocol and Provocation Tests

6.1 Standard Protocols for Medical Thermography

The following protocol is based on the clinical practice of recording thermal images of the human body continuously modified over 40 years [1–8].

6.1.1 Patient Preparation

Thermal imaging of the human body skin surface requires a set of standardized conditions. There are many factors and influences on the skin that will determine the actual temperature distribution that is under investigation. Environmental conditions and levels of physical activity are both strong influences. Therefore, it is common practice to allow the subject to be suitably prepared before commencing the image acquisition.

The subject should remove clothing from the areas of the body to be examined and should remain in a place such as a cubicle which is at a stable temperature and humidity. Vigorous exercise should normally be avoided prior to this period of stabilization.

The Thermal Human Body: A Practical Guide to Thermal Imaging
Kurt Ammer and Francis Ring

ISBN 978-981-4745-82-6 (Hardcover), 978-0-429-01998-2 (eBook)
www.jennystanford.com

Furthermore, smoking, large food intake and the use of topical medications should be avoided, at least for 2 h prior to test.

It is always recommended that the subjects should be informed preferably in writing of this requirement well in advance of their appointment, and the nature of the tests should be explained in order to obtain maximum co-operation. Medical imaging often involves exposure to some form of energy or radiation. It should be made clear that infrared imaging is passive, and only detects natural heat emitted by the skin.

The preparatory stabilization period should be at the same environmental conditions as those to be used for the imaging procedure

6.1.2 Patient Positioning

As is all medical imaging protocols, there are particular recommendations in the way in which the subject to positioned for regional imaging. A series of common positions have been described by Ammer [7] that cover many routine investigations by thermal imaging. The criteria are that ever image taken should follow a detailed description of the position of that patient and the anatomical limits of the target area. Unless this is followed, the results obtained cannot be reproduced, and therefore cannot be used to measure change on further examinations [9]. This lack of standardization has led in the past to much criticism of the unreliability of the technique, when in fact the blame should be placed on poor technique on the part of the investigator!

A source of infinite variables in imaging the human body is introduced when the camera is placed at an undetermined angle to the skin surface. The target area should always be at 90° to the camera plane, one reason why in Chapter 4 a single pillar stand is recommended as a camera mount rather than a tripod with a ball and socket variable mount. The four main anatomical aspects of human body imaging are shown in Fig. 6.1.

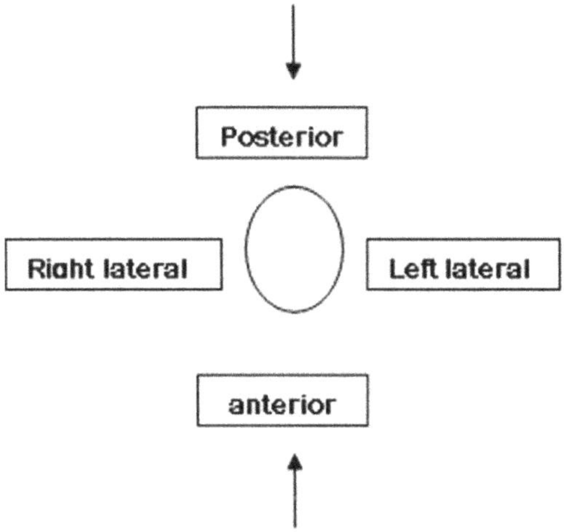

Figure 6.1 Four standard aspects for recording images of the human body surface.

6.1.3 Temperature of the Examination Room

A range of suitable room temperatures have been established, depending on the type of examination. When imaging the body trunk, or limbs and joints being investigated for inflammation, e.g. arthritis, a room temperature of 20°C is often recommended. However, when the extremities, hands and feet of lower limbs being investigated for changes in peripheral vascular or neurological abnormalities, a warmer ambient such as 22–24°C is preferable. Beyond these limits, in temperate climates, there is a risk of shivering below 17°C and sweating above 25°C. In hotter or colder climates, some differences in these temperatures will be found. This also refers to partially clothed subjects, which are the conditions normally found in clinical examinations. The totally unclothed body will be more subject to increased heat loss, and higher levels of ambient will be required to achieve a near stable skin temperature conditions.

However, whatever temperature is selected, variation for 1 h must be less than ±0.5°C.

6.1.4 Standard Views

When images are recorded of the human body, the size of the image is dependent on the distance between the camera and the subject. It is also dependant on the focal length of the camera lens. Many infrared cameras do not have the option for changing the lens, so it is good practice to maintain a constant distance between the subject and the camera lens to achieve a comparable field of view. As this also will influence the number of pixels and the spatial and thermal resolution in the image, repeated imaging over time of the same target area of the body can be affected by variables, if the procedure is not standardized.

A solution to rapid reproducibility has been devised at the University of Glamorgan, UK (now University of South Wales) [10]. Defined body positions and outline or mask on the image capture screen for that specific view has advantages. First, there are variations in size between subjects, so in this case the camera distance is adjusted so that the thermogram selected fits into the mask on every occasion. This has been tested in studies of reproducibility [11].

The following images show a documented set of standard views for thermal imaging that, as described, are an aid to reproducible thermograms.

Figure 6.2 Standard views of the human body (24 views)

1. Total body (anterior view); Code: TBA

Upper edge of the image: The most cranial point of the head	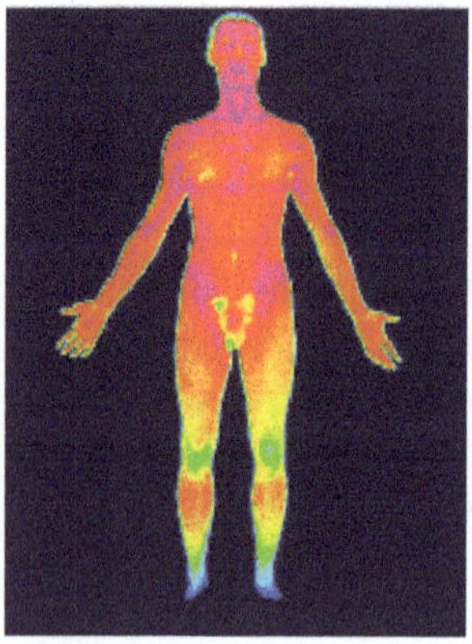
Lower edge of the image: Soles of the feet	
Other conditions Arms and legs slightly abducted, palms point forwards, head is a vertical position, neither rotated nor tilted to the side	

2. Total body (dorsal view); Code: TBD

Upper edge of the image: The most cranial point of the head	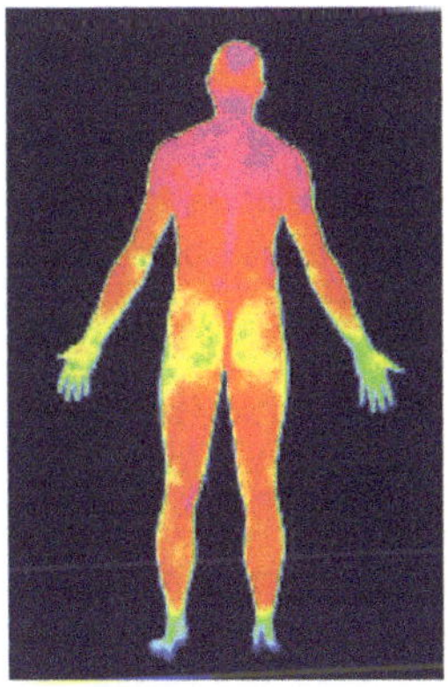
Lower edge of the image: Soles of the feet	
Other conditions Arms and legs slightly abducted, dorsum of the hands point towards the camera, head is a vertical position, neither rotated nor tilted to the side	

3. Total body (lateral view); Code: TBL (left) or TBR (right)

Upper edge of the image: The most cranial point of the head	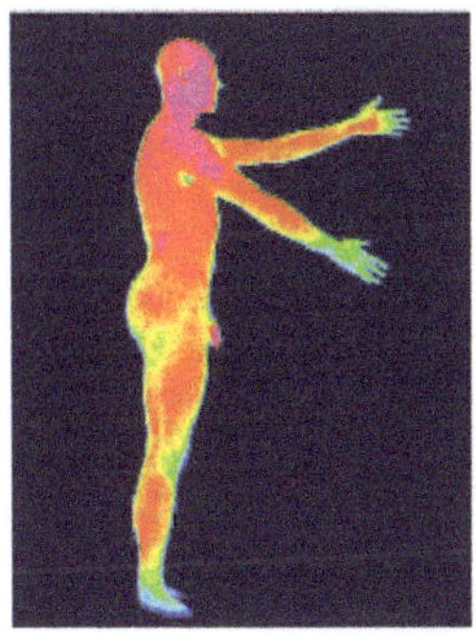
Lower edge of the image: Soles of the feet	
Other conditions: Arms scissored, left hand side of the points towards the camera, head is a vertical position, not rotated or tilted to the side	

BODY REGION

4. Face (anterior view); Code: FA

Upper edge of the image: The most cranial point of the head	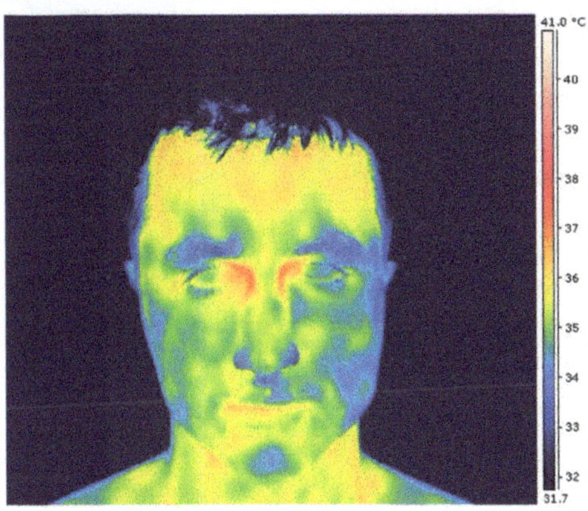
Lower edge of the image: Below the chin at the level of hyoid bone	
Other conditions Head is a vertical position, not rotated or tilted to the side	

4a. Upper Face (anterior view); Code: UFA

Lateral edges of the image:
Both ears are within the image

Lower edge of the image:
Below the nasal base

Other conditions
Head is a vertical position, not rotated or tilted to the side

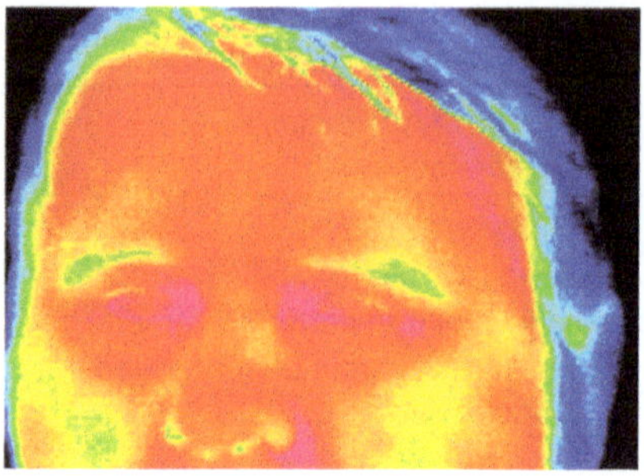

5. Neck (dorsal view); Code: ND

Upper edge of the image:
Line between the cranial ends of the ear

Lower edge of the image:
on the level of the processus spinosus of D1

Other conditions
Head is a vertical position, not rotated or tilted to the side

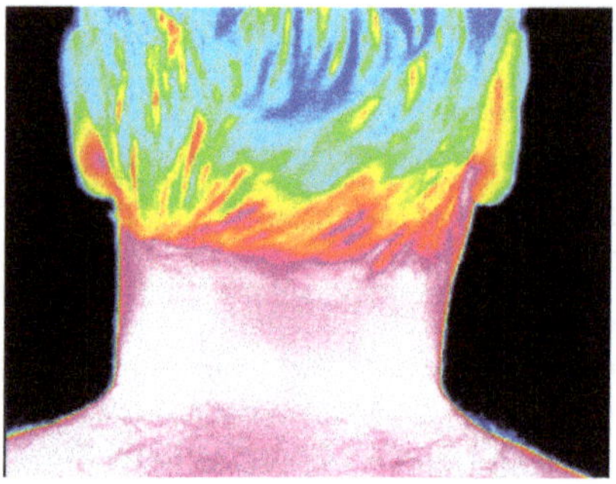

6. Chest (anterior view); Code: CA

Upper edge of the image:
Below the chin at the level of hyoid bone

Lower edge of the image:
Below the rib cage (approximately the tip of the elbow)

Other condition
Arms slightly abducted, the outline of both shoulders and upper arms are within the image

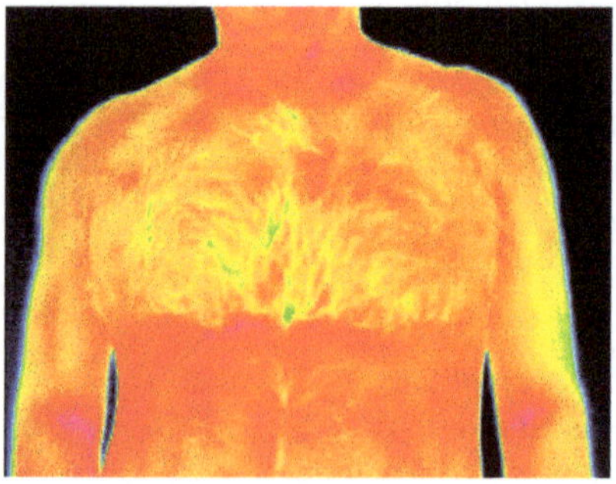

7. Upper Back; Code: UB

Upper edge of the image: Level of the processus spinosus of C6	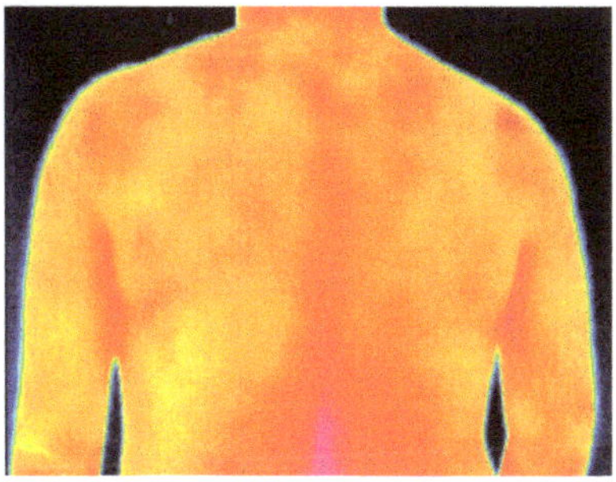
Lower edge of the image: Below the rib cage (approximately the tip of the elbow	
Other condition: Arms slightly abducted, the outline of both shoulders and upper arms are within the image	

8. Right arm (anterior view); Code: RAA

Upper edge of the image: Wrist	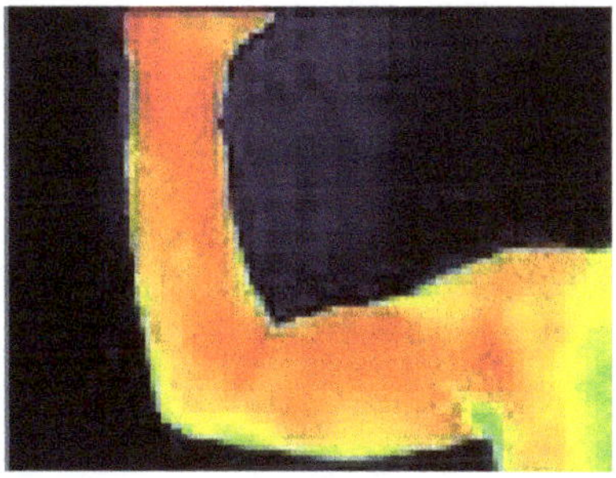
Lower edge of the image: Below the axillary fold	
Other condition: Arm 90° abducted and elbow 90° bent, the outline of the of the deltoid muscle is within the image	

9. Right arm (dorsal view); Code: RAD

Upper edge of the image: Wrist	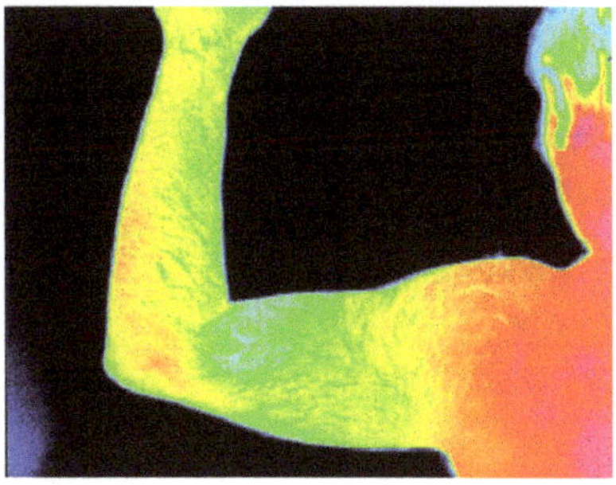
Lower edge of the image: Below the axillary fold	
Other condition Arm 90° abducted and elbow 90° bent, the outline of the of the deltoid muscle is within the image	

10. Left arm (anterior view); Code: LAA

Upper edge of the image: Wrist	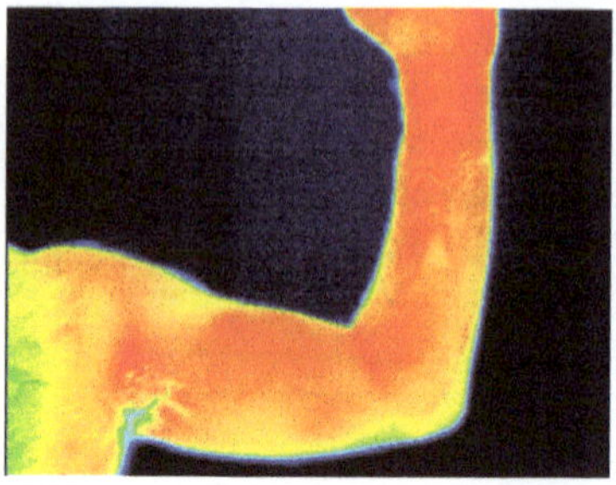
Lower edge of the image: Below the axillary fold	
Other condition Arm 90° abducted and elbow 90° bent, the outline of the of the deltoid muscle is within the image	

11. Left arm (dorsal view); Code: LAD

Upper edge of the image: Wrist	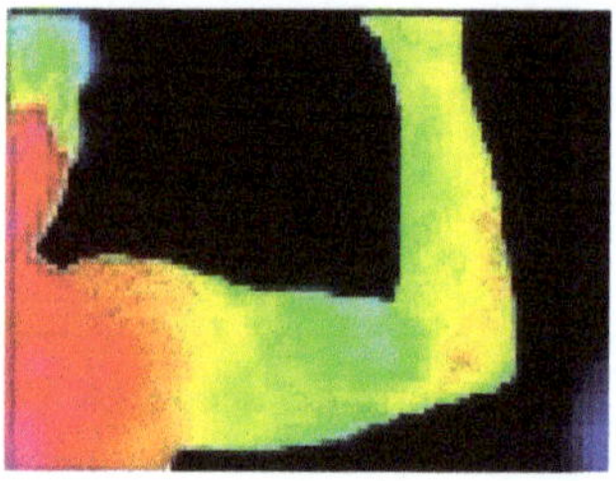
Lower edge of the image: Below the axillary fold	
Other condition: Arm 90° abducted and elbow 90° bent, the outline of the of the deltoid muscle is within the image	

12. Both Hands (dorsal view); Code: BHD

Upper edge of the image: Wrist	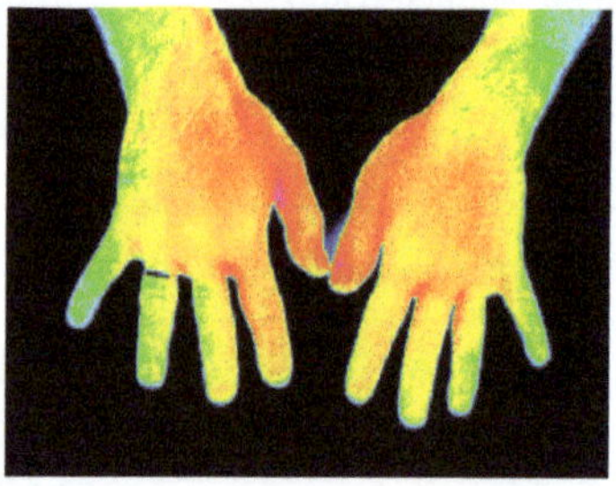
Lower edge of the image: Tip of the middle finger	
Other condition: Middle fingers are parallel, thumbs do not touch	

13. Abdomen; Code

Upper edge of the image: Anterior fold of the axilla	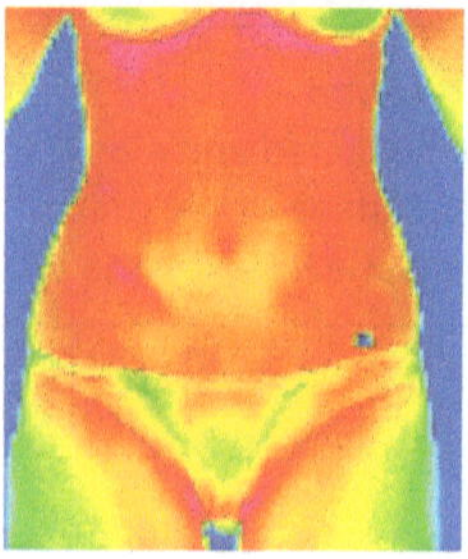
Lower edge of the image: Lower end of the groin	
Other condition: Arms and legs slightly abducted	

14. Lower Back; Code: LB

Upper edge of the image: Posterior fold of the axilla	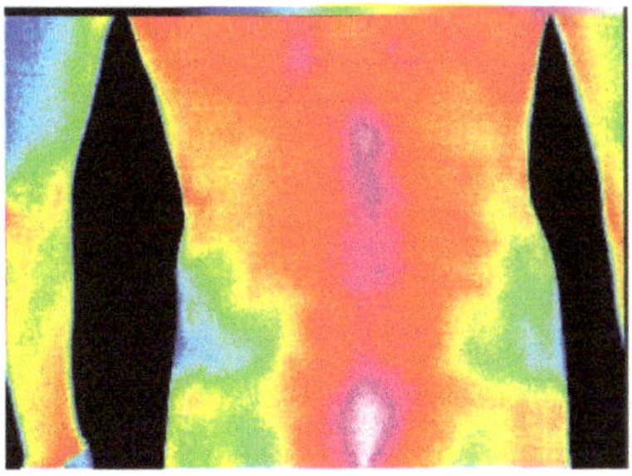
Lower edge of the image: Upper end of the natal cleft	
Other condition: Arms and legs slightly abducted	

15. Thighs (anterior view); Code: TA

Upper edge of the image: Line at the iliac crest	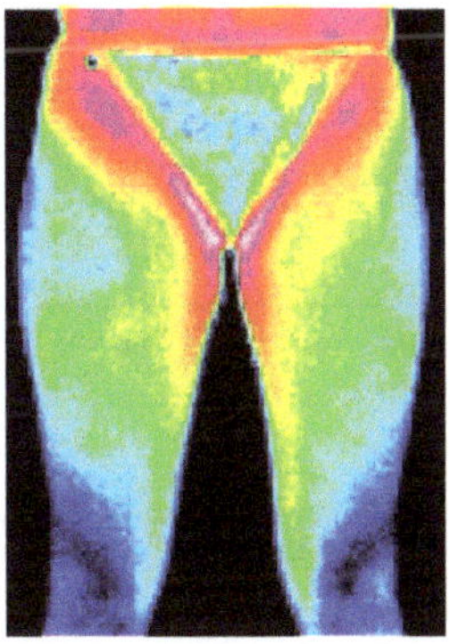
Lower edge of the image: Tip of the patella	
Other condition: Legs slightly abducted, feet parallel and the big toes point towards the camera	

16. Thighs (dorsal view); Code: TD

Upper edge of the image: Line at the iliac crest	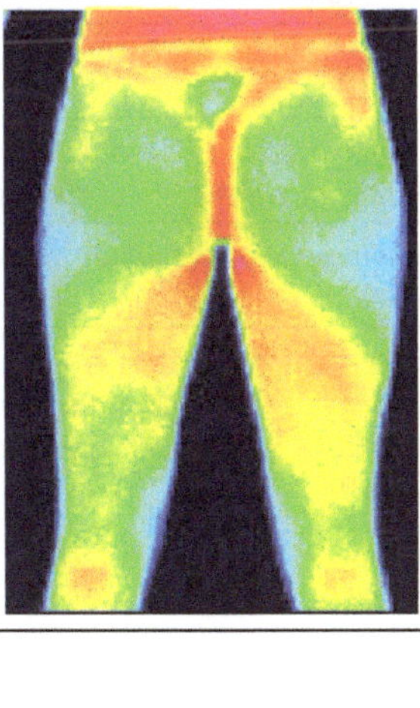
Lower edge of the image: Head of the fibula	
Other condition Legs slightly abducted, feet parallel and the big toes point towards the camera	

17 Leg, right (lateral view); Code: LRL

<table>
<tr><td>Upper edge of the image:
Iliac crest</td><td rowspan="3">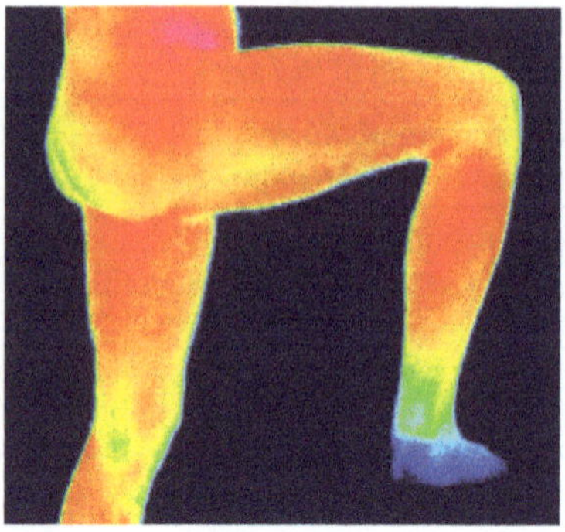</td></tr>
<tr><td>Lower edge of the image:
Sole</td></tr>
<tr><td>Other condition
Hip and knee are approximately 90° bent, foot is placed on a chair, the total leg is in the sagittal plane of the body</td></tr>
</table>

18. Leg, left (lateral view); Code: LLL

<table>
<tr><td>Upper edge of the image:
Iliac crest</td><td rowspan="3">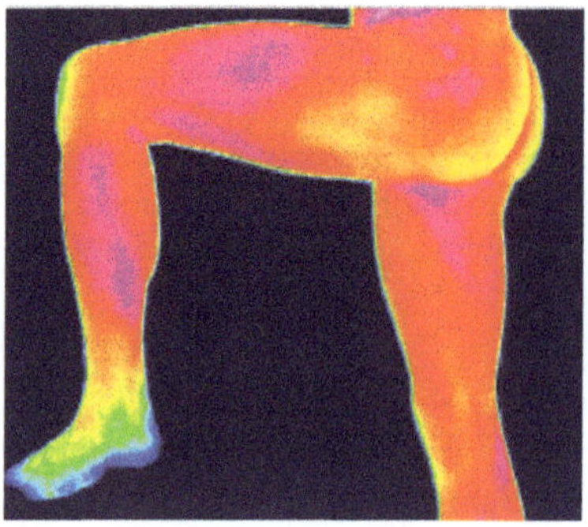</td></tr>
<tr><td>Lower edge of the image:
Sole</td></tr>
<tr><td>Other condition:
Hip and knee are approximately 90° bent, foot is placed on a chair, the total leg is in the sagittal plane of the body</td></tr>
</table>

19. Lower Legs (anterior view); Code: LLA

<table>
<tr><td>Upper edge of the image:
1 inch above the rim of the patella</td><td rowspan="3">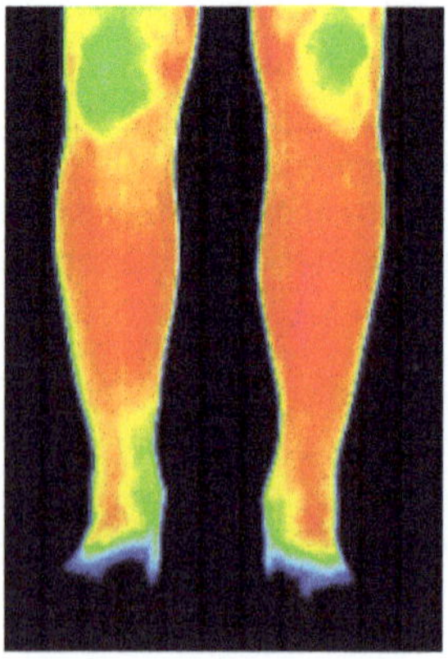</td></tr>
<tr><td>Lower edge of the image:
Soles</td></tr>
<tr><td>Other condition:
Legs slightly abducted, feet parallel and the big toes point towards the camera</td></tr>
</table>

20. Lower Legs (dorsal view); Code: LLD

<table>
<tr><td>Upper edge of the image:
1 inch above the femoral condyli</td><td rowspan="3">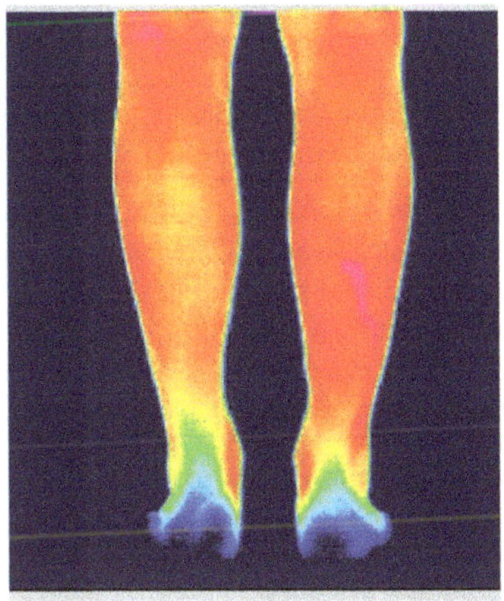</td></tr>
<tr><td>Lower edge of the image:
Soles</td></tr>
<tr><td>Other condition:
Legs slightly abducted, feet parallel and the big toes point towards the camera</td></tr>
</table>

21. Both Knees (anterior view); Code: BKA

<table>
<tr><td>Upper edge of the image:
1.5 inch above the rim of the patella</td><td rowspan="3">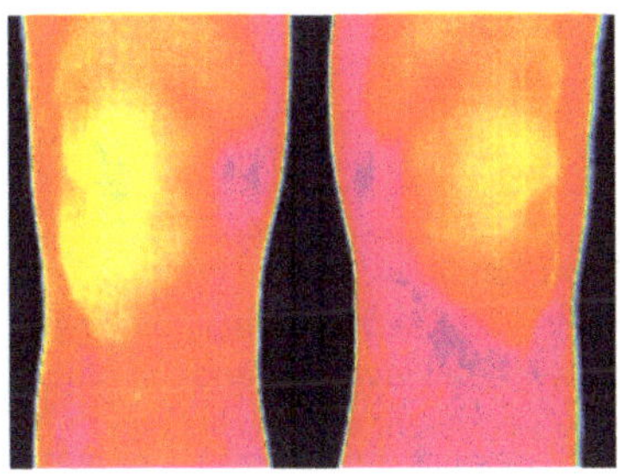</td></tr>
<tr><td>Lower edge of the image:
1 inch below the tip of the fibulae</td></tr>
<tr><td>Other condition:
Legs slightly abducted, feet parallel and the big toes point towards the camera</td></tr>
</table>

22. Both Ankles (anterior view); Code: BAA

<table>
<tr><td>Upper edge of the image:
1 inch above the malleoli</td><td rowspan="3">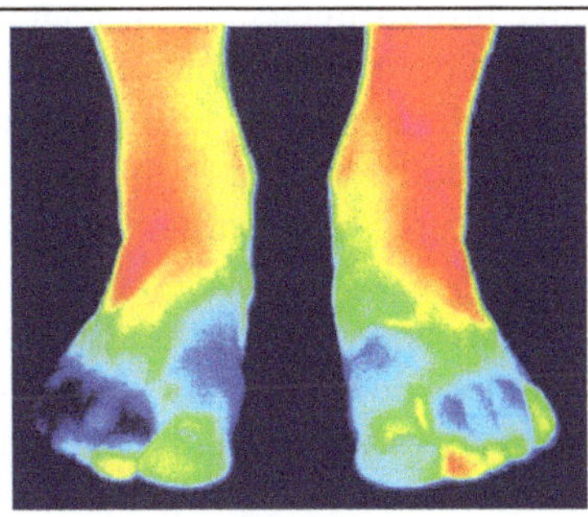</td></tr>
<tr><td>Lower edge of the image:
Soles</td></tr>
<tr><td>Other condition:
Legs slightly abducted, feet parallel and the big toes point towards the camera</td></tr>
</table>

23. Dorsal Feet; Code: DF

Upper edge of the image: Ankle	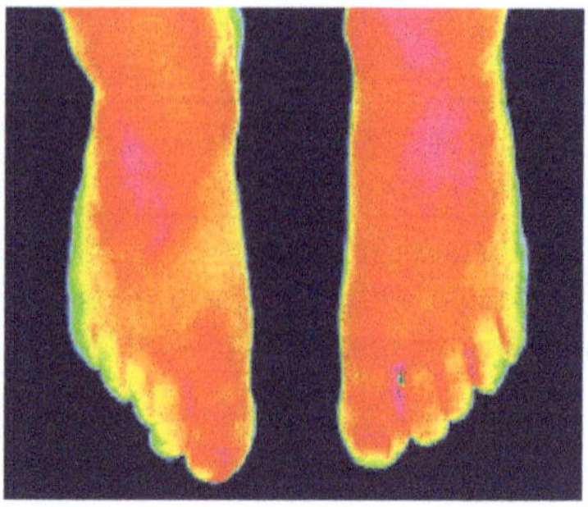
Lower edge of the image: Tip of the great toe	
Other condition: Subject sits on a chair, legs slightly abducted, feet parallel and the big toes point forwards Camera is in a rectangle position to the dorsal surface of the feet	

24. Plantar Feet; Code: PF

Upper edge of the image: Tip of the great toe	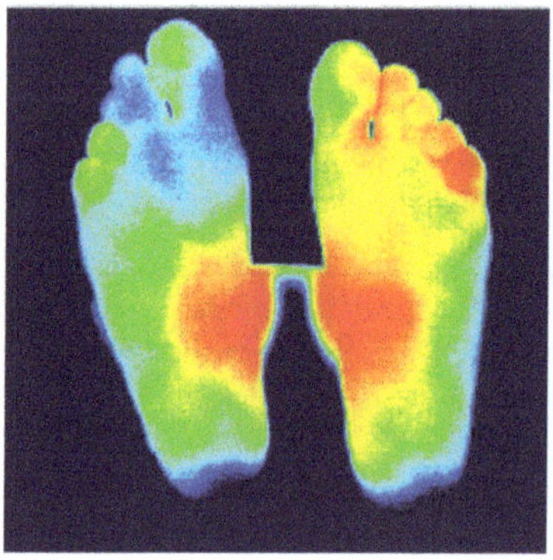
Lower edge of the image: Heel	
Other condition: Heels are placed on a chair, the angle between the lower leg and the foot is 90°, both feet are parallel	

6.1.5 Regions of interest

Image analysis is now carried out by computer, using powerful image processing techniques. The principle requirement is to define a region of interest from which the digital data can be extracted. In thermal imaging, this translates to temperature, which may be in a specific point or area in the image. Maximum, minimum and mean temperatures are commonly obtained from the defined region of interest.

It has been shown that the delineation of the region if interest is often made in a purely subjective manner, and therefore another source of variability [11]. By using anatomical landmarks for the parameters of every region of interest, a considerably improved reproducibility can be obtained. Furthermore, when another operator is introduced to perform the same task, a much closer match in results can be obtained using this method.

Figure 6.3 Regions of interest (ROI). A total of 90 ROIs can be defined as shown.

1. View: Total body (anterior); Code: TBA

Number of ROIs: 1	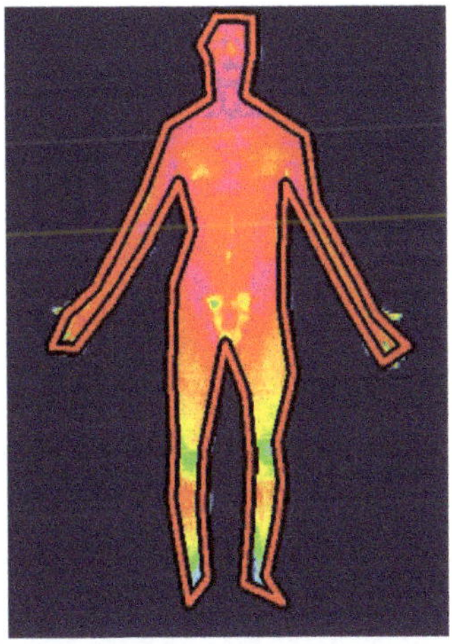
Shape*:* polygon, following the outline of the body	

2. View: Total body (dorsal); Code: TBD

Number of ROIs: 1	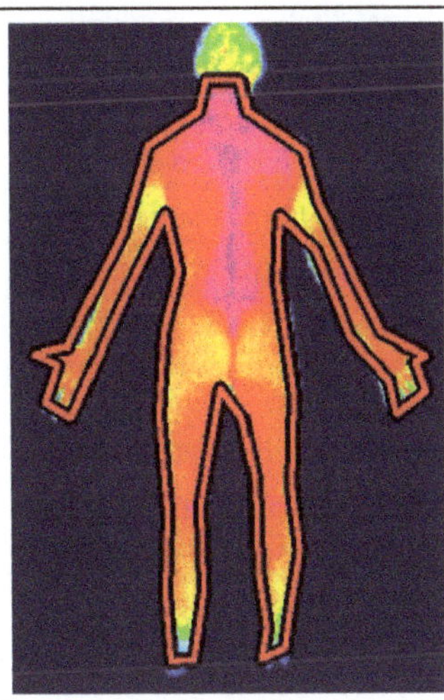
Shape*:* polygon, following the outline of the body	

3. View: Total body (lateral); Code: TBL (left) or TBR (right)

Number of ROIs: 1	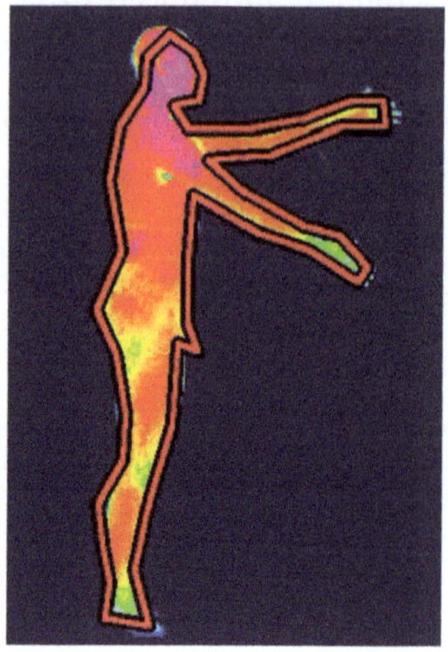
Shape: polygon, following the outline of the body	

4. View: Face; Code: FA

Number of ROIs: 4	
ROI 1: half of the face (red)	
Shape: polygon	
Outline of the face from the scalp to the chin, following the midline of the face	
ROI 2: half of the face (blue)	
Shape: polygon	
Outline of the face from the scalp to the chin, following the midline of the face	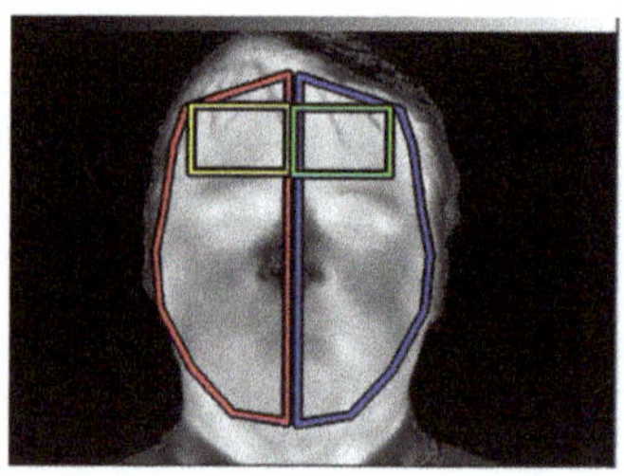
ROI 3: half of the forehead (yellow)	
Shape: rectangle	
Lower edge: adjacent to the right eyebrow **Right upper corner**: adjacent to the midline of the face **Left upper corner:** adjacent to the hair line	
ROI 4: half of the forehead (green)	
Shape: rectangle	
Lower edge: adjacent to the left eyebrow **Right upper corner**: adjacent to the hair line **Left upper corner:** adjacent to the midline of the face	

5. View. Upper Face (anterior view); Code: UFA

Number of ROIs: 2	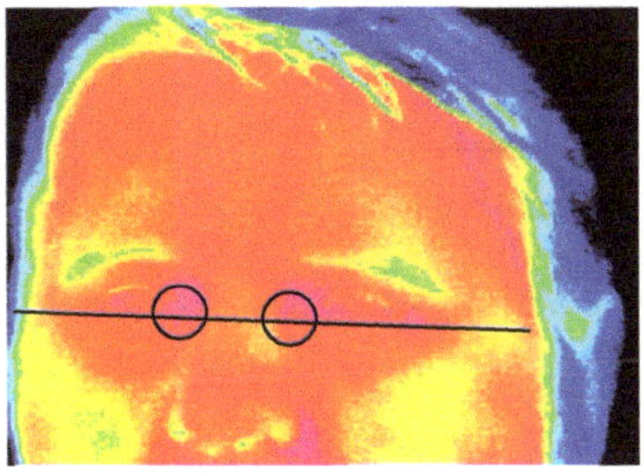
ROI 1: right inner canthus	
Shape: Circle	
Connect the outer canthus of the eyes with a straight line. Define a circle with the diameter equal to the distance between the inner canthus and the medial outline of the orbita. The circle is aligned to the outline of the orbita and the connection line crosses the circle at 1/3 radius below the centre.	
ROI 2: left inner canthus	
Shape: Circle	

6. View: Neck: Code: ND

Number of ROIs: 2	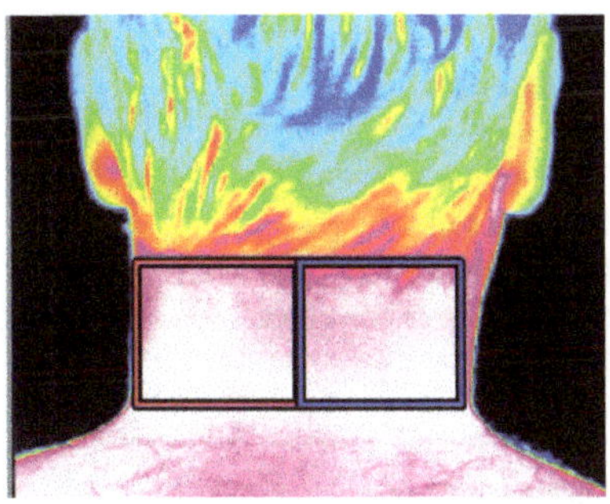
ROI 1: half of the neck (red)	
Shape: rectangle	
Upper edge: adjacent to the hair line **Lower right corner**: adjacent to the junction of the outline of the trapeze muscle with the outline of the neck **Left edge:** adjacent to midline of the neck	
ROI 2: half of the neck (blue)	
Shape: rectangle	
Upper edge: adjacent to the hair line **Right edge:** adjacent to midline of the neck **Lower left corner**: adjacent to the junction of the outline of the trapeze muscle with the outline of the neck	

7. View: Chest; Code: CA

<table>
<tr><td>Number of ROIs: 4</td><td rowspan="13">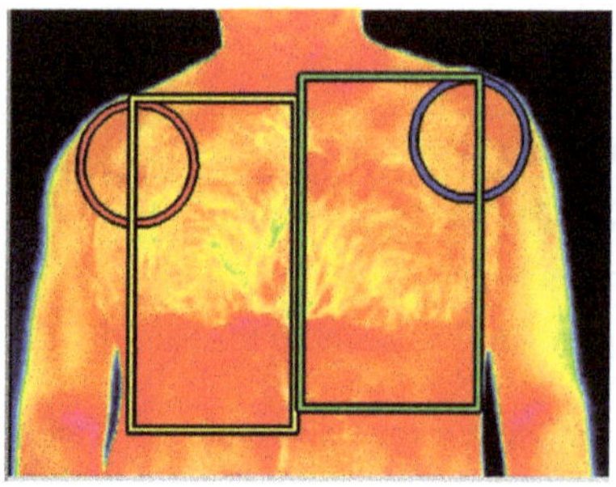</td></tr>
<tr><td>ROI 1: right anterior shoulder joint (red)</td></tr>
<tr><td>Shape: circle</td></tr>
<tr><td>Outline of the circle is adjacent to the acromion and also to the anterior axillary fold</td></tr>
<tr><td>ROI 2: left anterior shoulder joint (blue)</td></tr>
<tr><td>Shape: Circle</td></tr>
<tr><td>Outline of the circle is adjacent to the acromion and also to the anterior axillary fold</td></tr>
<tr><td>ROI 3: half of the chest (yellow)</td></tr>
<tr><td>Shape: rectangle</td></tr>
<tr><td>Lower edge: aligned with the right elbow
Right upper corner: adjacent to the acromion
Left upper corner: adjacent to the midline of the chest</td></tr>
<tr><td>ROI 4: half of the chest (green)</td></tr>
<tr><td>Shape: rectangle</td></tr>
<tr><td>Lower edge: aligned with the left elbow
Right upper corner: adjacent to the midline of the chest
Left upper corner: adjacent to the acromion</td></tr>
</table>

8. View: Upper Back: Code: UB

<table>
<tr><td>Number of ROIs: 4</td><td rowspan="12">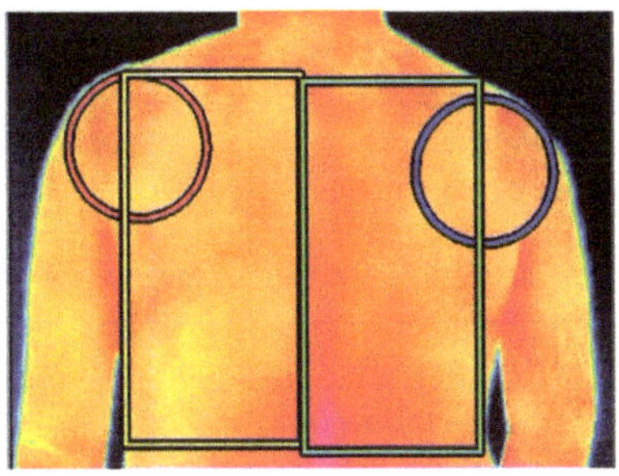</td></tr>
<tr><td>ROI 1: left posterior shoulder joint (red)</td></tr>
<tr><td>Shape: circle</td></tr>
<tr><td>Outline of the circle is adjacent to the acromion and also to the posterior axillary fold</td></tr>
<tr><td>ROI 2: right anterior shoulder joint (blue)</td></tr>
<tr><td>Shape: Circle</td></tr>
<tr><td>Outline of the circle is adjacent to the acromion and also to the posterior axillary fold</td></tr>
<tr><td>ROI 3: half of the upper back (yellow)</td></tr>
<tr><td>Shape: rectangle</td></tr>
<tr><td>Lower edge: aligned with the left elbow
Right upper corner: adjacent to the acromion
Left upper corner: adjacent to the midline of the chest</td></tr>
<tr><td>ROI 4: half of the upper back (green)</td></tr>
<tr><td>Shape: rectangle</td></tr>
<tr><td>Lower edge: aligned with the right elbow
Right upper corner: adjacent to the midline of the chest
Left upper corner: adjacent to the acromion</td><td></td></tr>
</table>

9. Right arm (anterior view); Code: RAA

Number of ROIs: 3	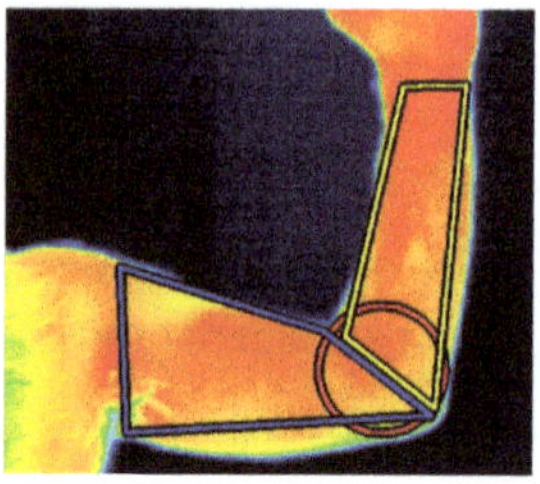
ROI 1: medial elbow (red)	
Shape: circle	
Outline of the circle is adjacent to the cubital fold and the lower edges of the elbow	
ROI 2: anterior upper arm (blue)	
Shape: trapezoid	
Upper left corner: insertion of the deltoid muscle **Upper right corner**: cubital fold **Lower right corner:** tip of the elbow **Lower left corner:** axillary fold	
ROI 3: anterior forearm (yellow)	
Shape: trapezoid	
Upper left corner: radial end of wrist **Upper right corner:** ulnar end of the wrist **Lower right corner:** tip of the elbow **Lower left corner:** cubital fold	

10. Left arm (anterior view); Code: LAA

Number of ROIs: 3	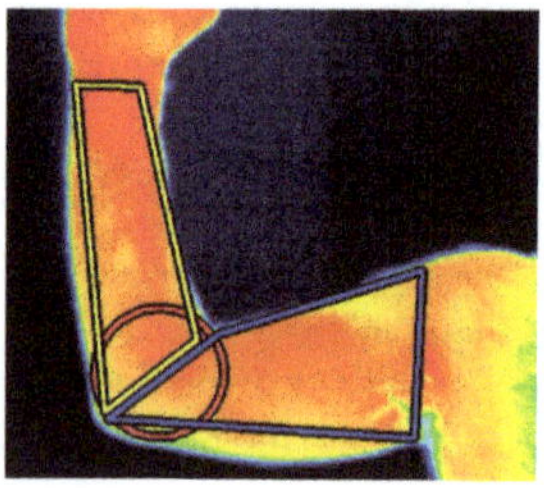
ROI 1: medial elbow (red)	
Shape: circle	
Outline of the circle is adjacent to the cubital fold and the lower edges of the elbow	
ROI 2: anterior upper arm (blue)	
Shape: trapezoid	
Upper left corner: cubital fold **Upper right corner:** insertion of the deltoid muscle **Lower right corner:** axillary fold **Lower left corner**: tip of the elbow	
ROI 3: anterior forearm (yellow)	
Shape: trapezoid	
Upper left corner: ulnar end of the wrist **Upper right corner:** radial end of the wrist **Lower right corner:** cubital fold **Lower left corner:** tip of the elbow	

11. Right arm (dorsal view); Code: RAD

Number of ROIs: 3	
ROI 1: lateral elbow (red)	
Shape: circle	
Outline of the circle is adjacent to the cubital fold and the lower edges of the elbow	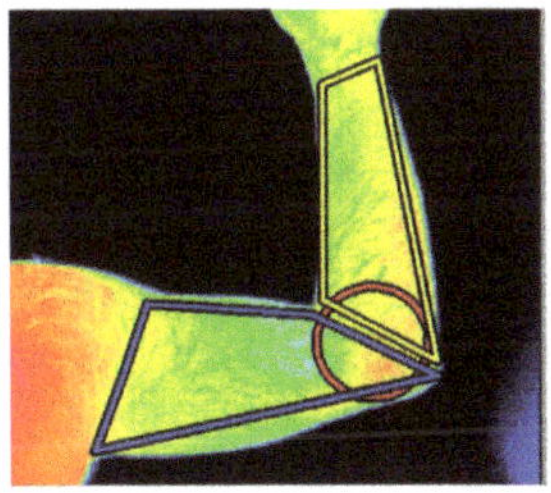
ROI 2: posterior upper arm (blue)	
Shape: trapezoid	
Upper left corner: insertion of the deltoid muscle **Upper right corner:** cubital fold **Lower right corner:** tip of the elbow **Lower left corner:** axillary fold	
ROI 3: posterior forearm (yellow)	
Shape: trapezoid	
Upper left corner: radial end of the wrist **Upper right corner:** ulnar end of the wrist **Lower right corner**: tip of the elbow **Lower left corner:** cubital fold	

12. Left arm (posterior view); Code: LAD

Number of ROIs: 3	
ROI 1: lateral elbow (red)	
Shape: circle	
Outline of the circle is adjacent to the cubital fold and the lower edges of the elbow	
ROI 2: posterior upper arm (blue)	
Shape: trapezoid	
Upper left corner: cubital fold **Upper right corner:** insertion of the deltoid muscle **Lower right corner:** axillary fold **Lower left corner:** tip of elbow	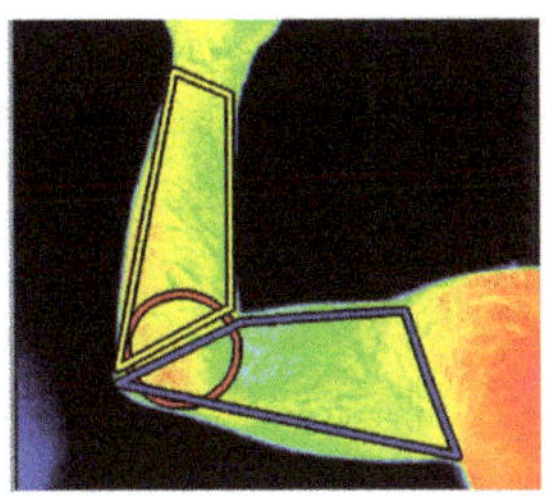
ROI 3: posterior forearm (yellow)	
Shape: trapezoid	
Upper left corner: ulnar end of the wrist **Upper right corner:** radial end of the wrist **Lower right corner:** cubital fold **Lower left corner: tip of the elbow**	

13. View: Both Hands (dorsal view); Code: BHD

Number of ROIs: 32, 1 over each wrist, 3 on each finger **X-sections**: 10, 1 on each finger

Wrist	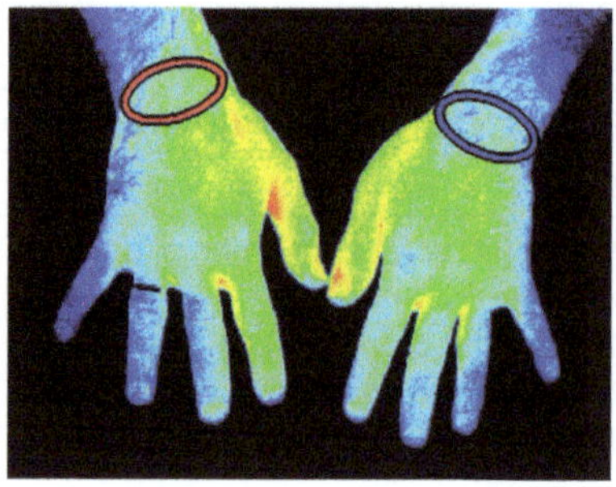
ROI 1 right Wrist (red)	
Shape: ellipse	
Adjacent to the lateral edges of the wrist, proportion of both axes of the ellipse is approximately 1:3	
ROI 2: half of the neck (blue)	
Shape: ellipse	
Adjacent to the lateral edges of the wrist, proportion of both axes of the ellipse is approximately 1:3	

Right thumb	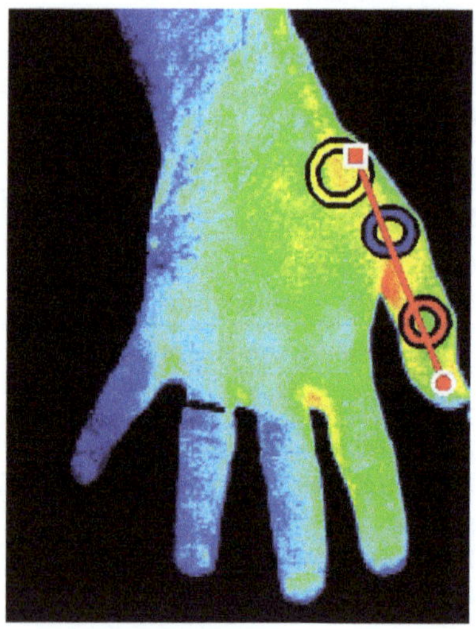
ROI 1: interphalangeal joint (red)	
Shape: circle	
Outline of the circle is adjacent to the edges of the interphalangeal joint	
ROI 2: metacarpophalangeal joint (blue)	
Shape: circle	
Outline of the circle is adjacent to the edges of the metacarpophalangeal joint	
ROI 3: carpo-metacarpal joint (yellow)	
Shape: circle	
Outline of the circle is adjacent to the edges of the carpo-metacarpal joint	
X-section (red) In the midline of the thumb from the fingertip to the proximal end of the metacarpus	

Left Thumb
ROI 1: interphalangeal joint (red)
Shape: circle
Outline of the circle is adjacent to the edges of the interphalangeal joint
ROI 2: metacarpophalangeal joint (blue)
Shape: circle
Outline of the circle is adjacent to the edges of the metacarpophalangeal joint
ROI 3: carpo-metacarpal joint (yellow)
Shape: circle
Outline of the circle is adjacent to the edges of the carpo-metacarpal joint
X-section (red)
In the midline of the thumb from the fingertip to the proximal end of the metacarpus

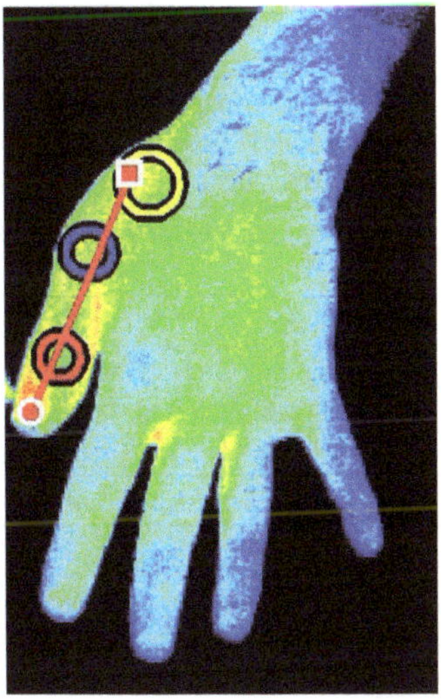

Right index finger
ROI 1: distal interphalangeal joint (red)
Shape: ellipse
Outline of the circle is adjacent to the fingertip and includes the proximal portion of the distal interphalangeal joint
ROI 2: proximal interphalangeal joint (blue)
Shape: circle
Outline of the circle is adjacent to the edges of the proximal interphalangeal joint
ROI 3: carpo-metacarpal joint (yellow)
Shape: circle
Outline of the circle is adjacent to the edges of the carpo-metacarpal joint
X-section (red)
In the midline of the index finger from the fingertip to the proximal end of the metacarpus

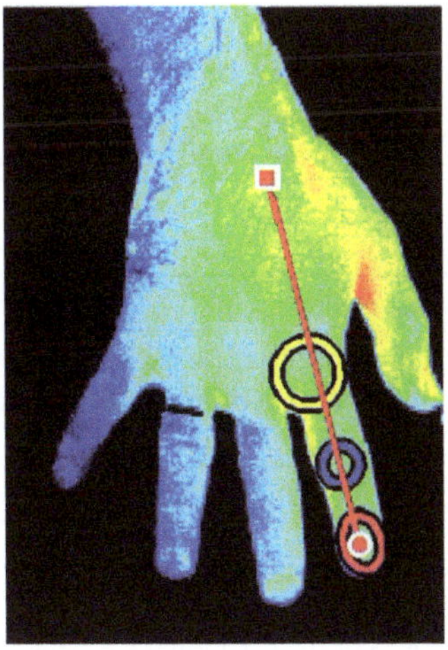

Left index finger	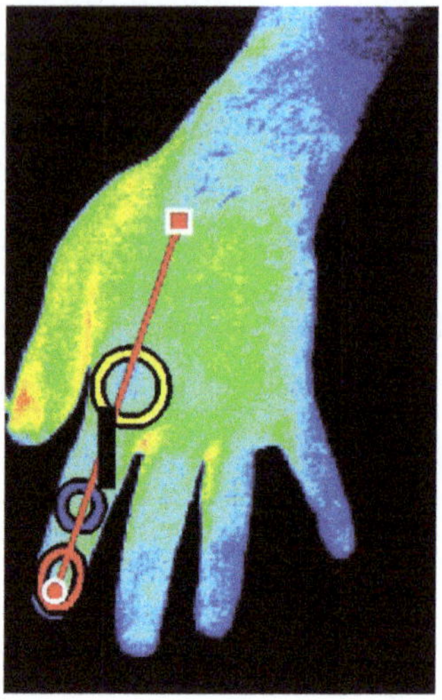
ROI 1: distal interphalangeal joint (red)	
Shape: ellipse	
Outline of the ellipse is adjacent to the fingertip and includes the proximal portion of the distal interphalangeal joint	
ROI 2: proximal interphalangeal joint (blue)	
Shape: circle	
Outline of the circle is adjacent to the edges of the proximal interphalangeal joint	
ROI 3: carpo-metacarpal joint (yellow	
Shape: circle	
Outline of the circle is adjacent to the edges of the carpo-metacarpal joint	
X-section (red) In the midline of the index finger from the fingertip to the proximal end of the metacarpus	

Right middle finger	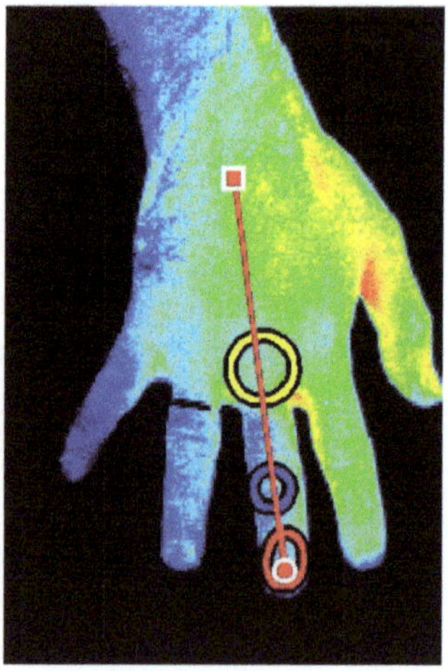
ROI 1: distal interphalangeal joint (red)	
Shape: ellipse	
Outline of the circle is adjacent to the fingertip and includes the proximal portion of the distal interphalangeal joint	
ROI 2: proximal interphalangeal joint (blue)	
Shape: circle	
Outline of the circle is adjacent to the edges of the proximal interphalangeal joint	
ROI 3: carpo-metacarpal joint (yellow)	
Shape: circle	
Outline of the circle is adjacent to the edges of the carpo-metacarpal joint	
X-section (red) In the midline of the middle finger from the fingertip to the proximal end of the metacarpus	

Left middle finger
ROI 1: distal interphalangeal joint (red)
Shape: ellipse
Outline of the circle is adjacent to the fingertip and includes the proximal portion of the distal interphalangeal joint
ROI 2: proximal interphalangeal joint (blue)
Shape: circle
Outline of the circle is adjacent to the edges of the proximal interphalangeal joint
ROI 3: carpo-metacarpal joint (yellow)
Shape: circle
Outline of the circle is adjacent to the edges of the carpo-metacarpal joint
X-section (red) In the midline of the middle finger from the fingertip to the proximal end of the metacarpus

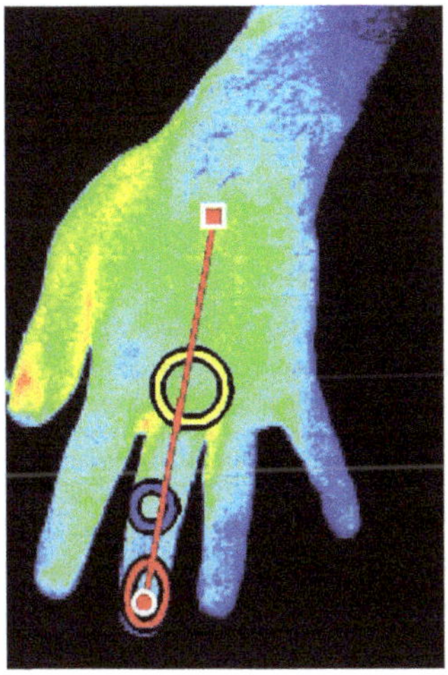

Right ring finger
ROI 1: distal interphalangeal joint (red)
Shape: ellipse
Outline of the circle is adjacent to the fingertip and includes the proximal portion of the distal interphalangeal joint
ROI 2: proximal interphalangeal joint (blue)
Shape: circle
Outline of the circle is adjacent to the edges of the proximal interphalangeal joint
ROI 3: carpo-metacarpal joint (yellow)
Shape: circle
Outline of the circle is adjacent to the edges of the carpo-metacarpal joint
X-section (red) In the midline of the ring finger from the fingertip to the proximal end of the metacarpus

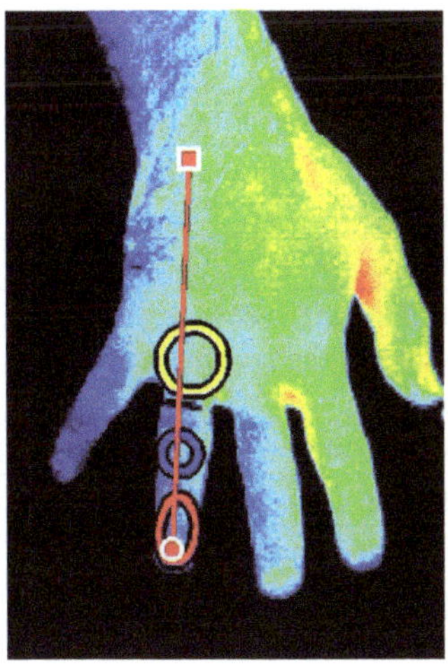

Left ring finger	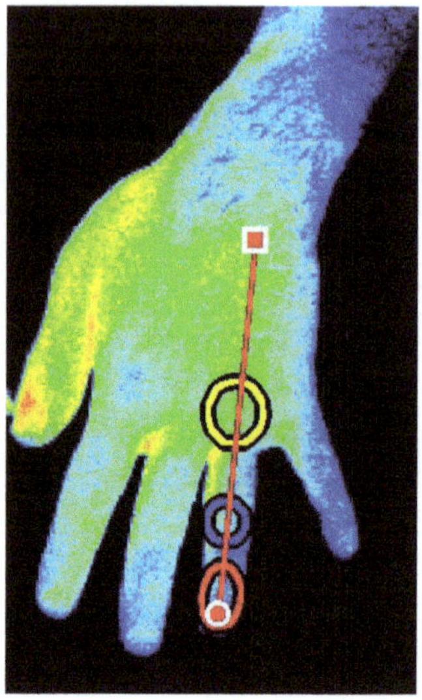
ROI 1: distal interphalangeal joint (red)	
Shape: ellipse	
Outline of the ellipse is adjacent to the fingertip and includes the proximal portion of the distal interphalangeal joint	
ROI 2: proximal interphalangeal joint (blue)	
Shape: circle	
Outline of the circle is adjacent to the edges of the proximal interphalangeal joint	
ROI 3: carpo-metacarpal joint (yellow)	
Shape: circle	
Outline of the circle is adjacent to the edges of the carpo-metacarpal joint	
X-section (red)	
In the midline of the ring finger from the fingertip to the proximal end of the metacarpus	

Right little finger	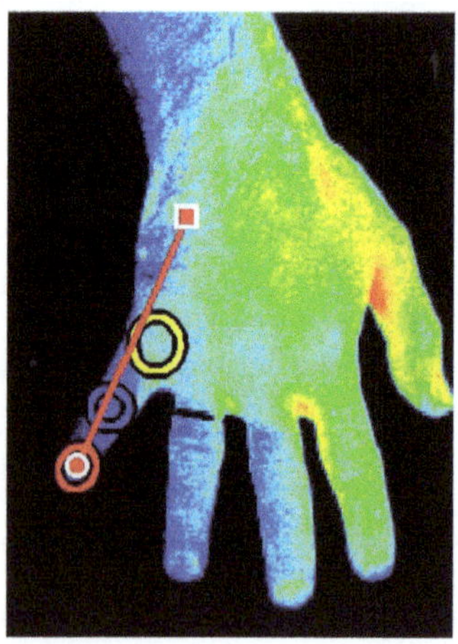
ROI 1: distal interphalangeal joint (red)	
Shape: ellipse	
Outline of the circle is adjacent to the fingertip and includes the proximal portion of the distal interphalangeal joint	
ROI 2: proximal interphalangeal joint (blue)	
Shape: circle	
Outline of the circle is adjacent to the edges of the proximal interphalangeal joint	
ROI 3: carpo-metacarpal joint (yellow)	
Shape: circle	
Outline of the circle is adjacent to the edges of the carpo-metacarpal joint	
X-section (red)	
In the midline of the little finger from the fingertip to the proximal end of the metacarpus	

Left little finger	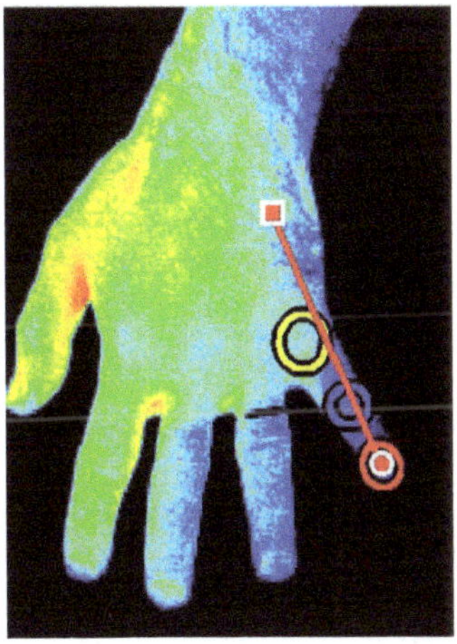
ROI 1: distal interphalangeal joint (red)	
Shape: circle	
Outline of the circle is adjacent to the fingertip and includes the proximal portion of the distal interphalangeal joint	
ROI 2: proximal interphalangeal joint (blue)	
Shape: circle	
Outline of the circle is adjacent to the edges of the proximal interphalangeal joint	
ROI 3: carpo-metacarpal joint (yellow)	
Shape: circle	
Outline of the circle is adjacent to the edges of the carpo-metacarpal joint	
X-section (red)	
In the midline of the little finger from the fingertip to the proximal end of the metacarpus	

14. View: Abdomen; Code: ABD

Number of ROIs: 2	
ROI 1: half of the abdomen (red)	
Shape: rectangle	
Upper left corner: aligned with the right elbow **Upper right corner:** at the midline of the body **Lower right corner:** lower end of the right groin **Lower left corner:** upper end of the right groin	
ROI 2: half of the abdomen (blue)	
Shape: rectangle	
Upper left corner: at the midline of the body **Upper right corner:** aligned with the left elbow **Lower right corner:** upper end of the left groin **Lower left corner:** lower end of the left groin	

15. View: Lower Back, Code: LB

Number of ROIs: 2	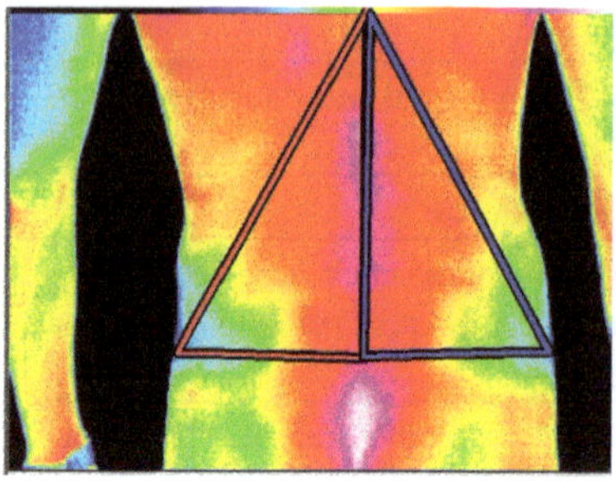
ROI 1: half of the lower back (red)	
Shape: triangle	
Upper corner: at the midline of the body **Left corner:** adjacent to the natal cleft **Right corner**: aligned with the natal cleft at the left side of the body	
ROI 2: half of the lower back (blue)	
Shape: rectangle	
Upper corner: at the midline of the body **Left corner:** aligned with the natal cleft at the left side of the body **Right corner**: adjacent to the natal cleft	

16. Thighs (anterior view); Code: TA

Number of ROIs: 2	
ROI 1: right thigh (red)	
Shape: polygon, following the outline of the thigh	
Upper left corner: upper end of the groin **Upper right corner:** lower end of the groin **Lower edge:** horizontal line aligned with the rim of the patella	
ROI 2: left thigh (blue)	
Shape: polygon, following the outline of the thigh	
Upper left corner: lower end of the groin **Upper right corner:** upper end of the groin **Lower edge:** horizontal line aligned with the rim of the patella	

17. Thighs (dorsal view); Code: T

Number of ROIs: 2	
ROI 1: left thigh (red)	
Shape: polygon, following the outline of the thigh	
Upper edge: horizontal line aligned with the gluteal fold **Lower edge:** horizontal line aligned with the tip of the fibula	
ROI 2: right thigh (blue)	
Shape: polygon, following the outline of the thigh	
Upper edge: horizontal line aligned with the gluteal fold **Lower edge:** horizontal line aligned with the tip of the fibula	

18. Leg, right (lateral view); Code: LRL

Number of ROIs: 3	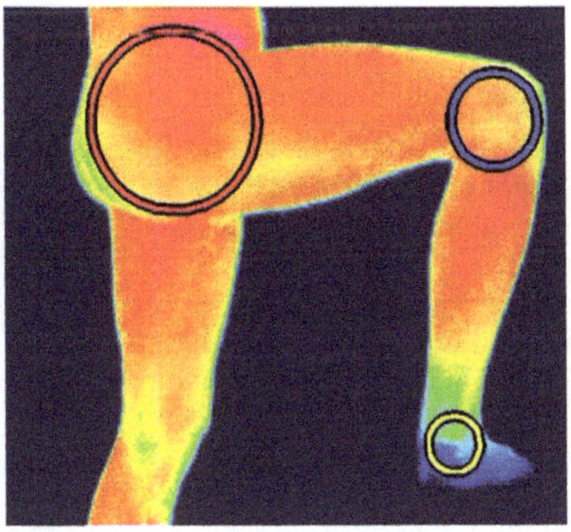
ROI 1: medial knee (red)	
Shape: circle	
Outline of the circle is adjacent to outline of the patella and the popliteal cubital fold	
ROI 2: lateral hip (blue)	
Shape: circle	
Outline of the circle is adjacent to outline of the nates	
ROI 3: lateral ankle (yellow)	
Shape: circle	
Outline of the circle is within the outlines of the ankle region	

19. Leg, left (lateral view); Code: LLL

Number of ROIs: 3	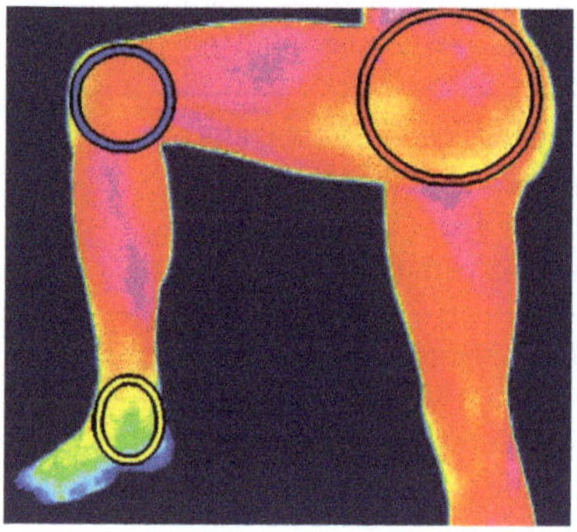
ROI 1: medial knee (red)	
Shape: circle	
Outline of the circle is adjacent to outline of the patella and the popliteal cubital fold	
ROI 2: lateral hip (blue)	
Shape: circle	
Outline of the circle is adjacent to outline of the nates	
ROI 3: lateral ankle (yellow)	
Shape: circle	
Outline of the circle is within the outlines of the ankle region	

20. Lower Legs (anterior view); Code: LLA

Number of ROIs: 2	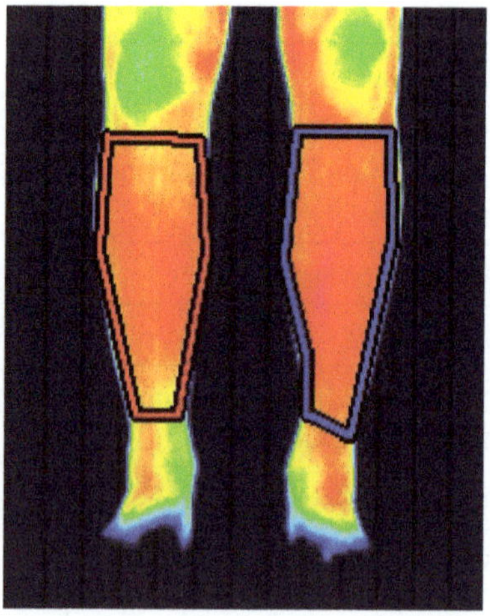
ROI 1: right lower leg (red)	
Shape: polygon, following the outline of the lower leg	
Upper edge: horizontal line aligned with the tip of the fibula **Lower edge:** horizontal line at the minimal width of the lower leg	
ROI 2: left lower leg (blue)	
Shape: polygon, following the outline of the lower leg	
Upper edge: horizontal line aligned with the tip of the fibula **Lower edge:** horizontal line at the minimal width of the lower leg	

21. Lower Legs (dorsal view); Code: LLD

Number of ROIs: 2	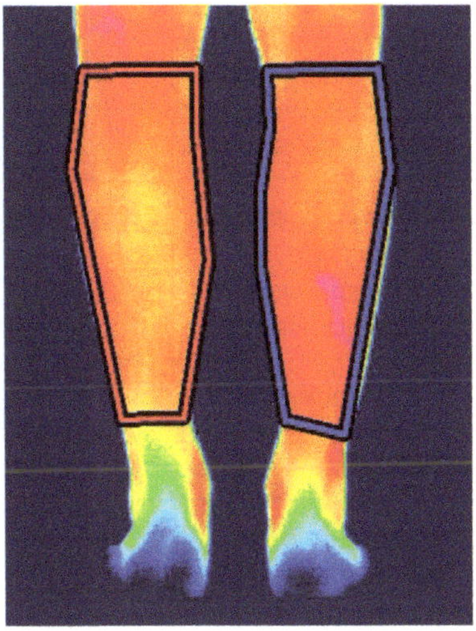
ROI 1: left lower leg (red)	
Shape: polygon, following the outline of the lower leg	
Upper edge: horizontal line aligned with the tip of the fibula **Lower edge:** horizontal line at the minimal width of the lower leg	
ROI 2: right lower leg (blue)	
Shape: polygon, following the outline of the lower leg	
Upper edge: horizontal line aligned with the tip of the fibula **Lower edge:** horizontal line at the minimal width of the lower leg	

22. Both Knees (anterior view); Code: BKA

Number of ROIs: 2	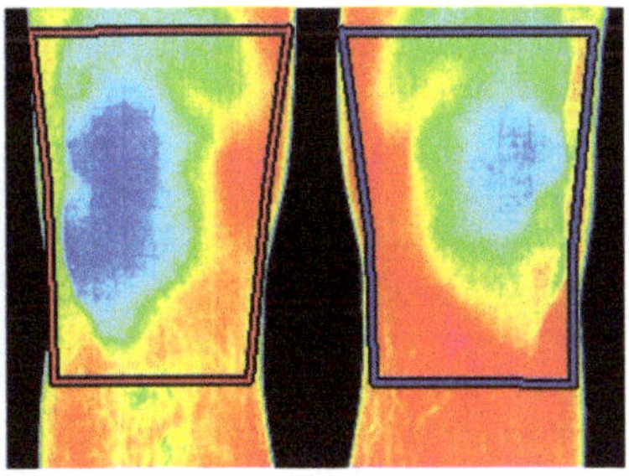
ROI 1: right knee (red)	
Shape: polygon following the outline of the knee	
Upper edge: horizontal line approximately 1 inch above the rim of the patella **Lower edge:** horizontal line aligned with the tip of the fibula	
ROI 2: left knee (blue)	
Shape: polygon following the outline of the knee	
Upper edge: horizontal line approximately 1 inch above the rim of the patella **Lower edge:** horizontal line aligned with the tip of the fibula	

23. Both Ankles (anterior view); Code: BAA

Number of ROIs: 2	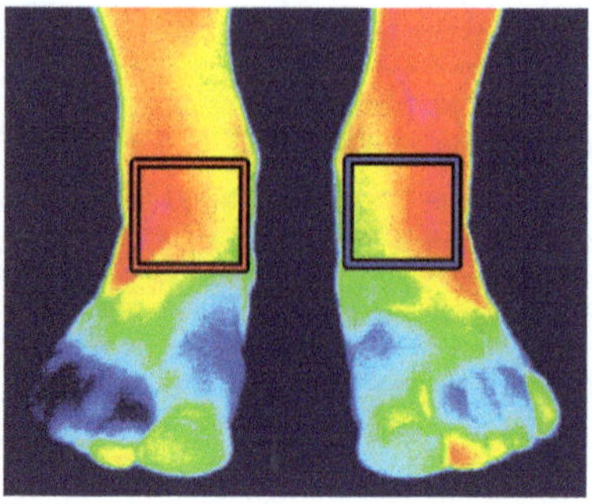
ROI 1: right ankle (red)	
Shape: square rectangle	
Upper edge: aligned with the tip of medial malleolus **Lower edge:** aligned with the tip of the naviculare bone	
ROI 2: left ankle (blue)	
Shape: square rectangle	
Upper edge: aligned with the tip of medial malleolus **Lower edge:** aligned with the tip of the naviculare bone	

24. Dorsal Feet; Code: DF

Number of ROIs: 2	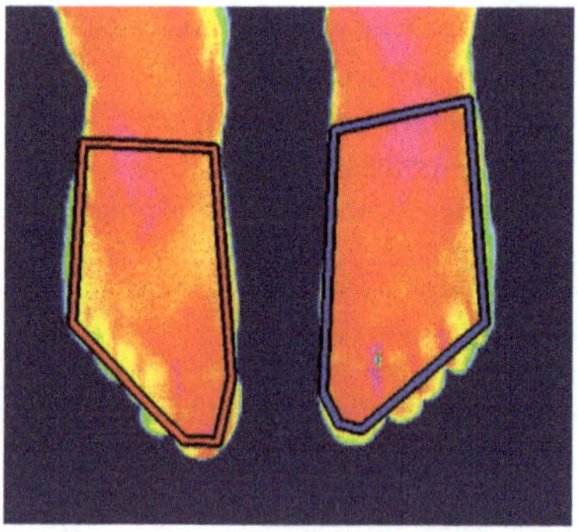
ROI 1: right foot (red)	
Shape: polygon, following the outline of the foot	
Upper edge: horizontal line aligned with the tip of the naviculare	
ROI 2: left ankle (blue)	
Shape: square rectangle	
Shape: polygon, following the outline of the foot	
Upper edge: horizontal line aligned with the tip of the naviculare	

25. Plantar Feet; Code: PF

Number of ROIs: 2	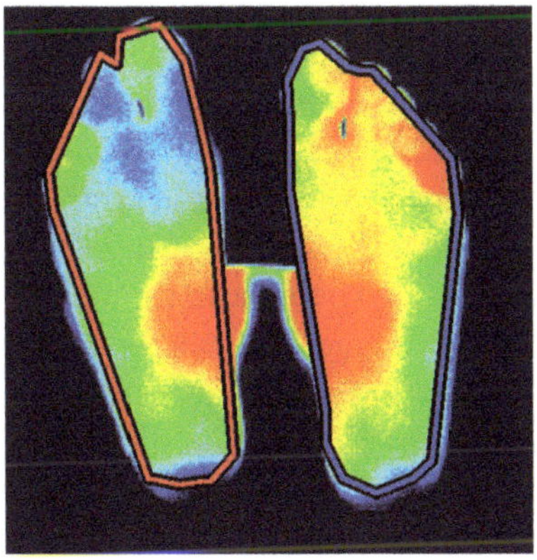
ROI 1: right sole (red)	
Shape: polygon, following the outline of the foot	
ROI 2: left sole (blue)	
Shape: polygon, following the outline of the foot	

6.2 Provocation Tests

Many different applications for thermal imaging in medicine have been described in the literature, most of which are based on expected thermal patterns after the period of thermal stabilization [12].

However, as a physiological imaging tool, the technique lends itself to non-invasive monitoring of skin temperature, which has altered status in response to an external stimulus. This can take the form of a thermal, mechanical, or drug induced stimulus.

In peripheral vascular abnormalities such as Raynaud's phenomenon of the hands, where fingers can become very cold with diminished peripheral blood flow, a thermal stress is often used. A commonly adopted practice is to record the thermogram of the dorsal hands after stabilization at room temperature of 22–24°C (temperate climates), then after covering both hands with thin plastic gloves, immersion in a water bath (often 20°C) for 1 min cools the whole hand by a few degrees. Further thermograms are recorded at time intervals from 10–20 min after cooling. A normal reaction to this stress is to find rapid recovery of finger temperatures, even reactive hyperaemia. In this case, the fingers become hotter than the dorsal hand for a few minutes before stabilizing back to even temperature distribution across the fingers and the dorsal hand area.

In clinical Raynaud's phenomenon, there is a different reaction. Recovery if any over the same time period will be slow, and colder fingers may persist for some time after testing [12].

Mechanical stress can take many forms and may be used to accentuate muscle weakness after exercising with heavy or bulky objects for a fixed time and rate.

Chemical stimuli, especially those used topically on the skin, can also indicate some localized neurological deficit. In brachial plexus injury for example it may be possible to show a lack of response to a vasodilator such as nicotinic acid spray on the skin, where the affected side fails to dilate and increase in temperature after application.

These tests require clinical knowledge in their interpretation, and mechanical or chemical loading must not be inappropriately applied by non-physicians, where injury could be inflicted, and legal proceedings might result.

6.2.1 Vibration Injury Testing

Work-related injuries are wide ranging, from problems with computer mouse users to machine and pneumatic hammer operators, etc. Originally described as repetitive strain injury (RSI), hand arm vibration syndrome (HAVS) occurs where the continual use of a machine or device can lead to injuries through handling over extended periods of time. In the upper limbs, there can be damage to muscles tendons joints and bones. Vibrating tools have developed in recent years from special tools for stone masons to vibrating spanners and drills for engineers and builders. Among a variety of tests that have been tried, thermal imaging offers a good passive monitoring technique that can indirectly indicate changes in vascular perfusion to the hands.

Vardasca has studied the requirements for a vibration stress that can be applied to the hands, and the effects of that stress can be measured by thermal imaging. A physical test of this kind also needs to be supported by a detailed questionnaire to ascertain the exposure conditions, and any documented symptoms. This approach shows promise but has yet to become an accepted means of testing as a national procedure to establish a clinical diagnosis. In the study by Vardasca, a vibration frequency of 31.5 Hz was used with an induced maximum acceleration of 36 m/s^2. The hand under test was placed in contact with the vibration plate for 2 min, and thermal imaging used to monitor the temperature distribution before and after exposure. Further studies in this area of industrial injury may lead to an established clinical test for occupational medicine [13].

6.2.2 Local Cooling for Detection of Vascular Patterns

In some applications, local vasculature is of interest to the surgeon, as in the case of operational surgery for varicose veins. The application of a local tourniquet to the legs with local cooling by cold sprays, cold packs, or just fan cooling can accentuate the perforating vessel sites sufficient for the skin marking to guide the surgeon. In breast reconstruction surgery, that may follow mastectomy, the surgeon needs to identify viable blood vessels to ensure good revascularization of the reconstructed tissue [14, 15].

References

1. Cosh JA, Ring EFJ. Thermography and rheumatology. *Rheumatology and Physical Medicine* 1970, 10(7), 342–348.
2. Collins AJ, Ring EFJ, Cosh JA, Bacon PA. l. Quantitation of thermography in arthritis using multi-isothermal analysis. *Annals of the Rheumatic Diseases* 1974, 33, 113–115.
3. Engel JM, Cosh JA, Ring EFJ, Page Thomas DP, Van Waes P, Shoenfeld D. Thermography in locomotor diseases - recommended procedure. *Journal of Rheumatology and Inflammation* 1979, 2, 299–306.
4. Ring EFJ. Standardisation of thermal imaging in medicine: Physical and environmental factors. In: Gautherie M, Albert E, Keith L, eds, *Thermal Assessment of Breast Health*, MTP Press Ltd. Lancaster/Boston/The Hague, 1983, pp. 29–36.
5. Ring EFJ. Quantitative thermal imaging. *Clinical Physics and Physiological Measurement.*. 1990, 11(Suppl A), 87–95.
6. Ring EFJ, Ammer K. The technique of infra red imaging in medicine. *Thermology International* 2000, 10(1), 7–14.
7. Ammer K. The Glamorgan Protocol for recording and evaluation of thermal images of the human body. *Thermology International* 2008, 18, 125–144.

8. Ring EFJ, Ammer K. Infrared thermal imaging in medicine. *Physiological Measurement* 2012, 33, R32–46.

9. Ammer K, Ring F. Standard procedures for infrared imaging in medicine. In: Diakides M, Bronzino JB, Peterson DR, eds, *Medical Infrared Imaging, Principles and Practices*, CRC Press, 2013, pp. 32.1–32.14.

10. Ammer K, Ring EFJ, Plassmann P, Jones BF. Rationale for standardised capture and analysis of infrared thermal images. In: Hutten H, Krösel P, eds, *Proceedings Part II, EMBEC'02 2.European Medical & Biological Engineering Conference*, IFMBE, Graz, 2002, pp. 1608–1609.

11. Ammer K, Ring EF. Repeatability of the standard view both dorsal hands. Results from a training course on medical infrared imaging. *Thermology International* 2004, 14(3), 100–103.

12. Ammer K. Cold challenge to provoke a vasospastic reaction in fingers determined by temperature measurements; a systematic review. *Thermology International* 2009, 19(4), 109–118.

13. Vardasca RAR. The effect of work related mechanical stress on the peripheral temperature of the hand, PhD Thesis. University of Glamorgan, UK, 2010.

14. de Weerd L, Weum S, Mercer JB. The value of dynamic infrared thermography (DIRT) in perforator selection and planning of free DIEP flaps. *Annals of Plastic Surgery*, 2009, 63(3), 274.

15. Mercer J, Miland AO. Dynamic infrared thermography in the assessment of skin perfusion. D.Sc University of Tromso, 2008.

Chapter 7

How Should a Clinical Thermal Image Be Recorded and Evaluated?

Medical thermal imaging is a service for physicians available on request. The referral to the infrared imaging service must include a written request which is the start of a chain of procedures, including a check of the equipment and the environment in which images will be recorded, preparation of the patients who should be imaged, selection of patient positions and body areas of interest, eventually application of provocation tests such as a mild cold challenge, and evaluation of the recorded images using sets of standardized measurement areas writing a report that includes all images and temperature values from the selected regions of interest.

7.1 Request for Thermal Imaging

One can only find significant changes of the skin temperature distribution in case of suspect of a disease. It is totally unacceptable to use a thermal imager for screening of diffuse and poorly defined complaints. It may therefore be helpful to provide a form for the referring physician in which the following minimum information should be reported (Fig. 7.1).

If the required information is not available, the thermographer must contact the referring physician and obtain the necessary information. Unless the thermographer is a qualified medical physician, the technician who conducts the infrared examination

The Thermal Human Body: A Practical Guide to Thermal Imaging
Kurt Ammer and Francis Ring

ISBN 978-981-4745-82-6 (Hardcover), 978-0-429-01998-2 (eBook)
www.jennystanford.com

must not try to find the reason for the requested examination. However, the thermographer should complete the information on symptoms in terms of actuality such as whether the symptoms presented at the moment or what time period passed since they were present last time. Also the dosage and kind of drugs and of physiotherapy may be added. All this information is part of the patient's record and must be documented with the conditions of image recording and its results.

Personal information of the patient:	
Required:	name gender □ male □ female date of birth
Optional:	occupation sports activity
Health problem	
Required:	Affected body region: .... Symptoms such as pain □ yes □ no paraesthesia □ yes □ no swelling □ yes □ no redness □ yes □ no heat sensation □ yes □ no muscle weakness □ yes □ no
Required	If a diagnosis was already established, what is the diagnosis?
Required	Recent and previous injuries in the symptomatic body region?
Required Optional	Recent surgery □ yes □ no if yes, what kind of operation?
Required	Current treatment such as Drugs □ yes □ no
Optional	if yes, which drugs, which dosage ....
Required	Physiotherapy □ yes □ no
Optional	if yes, which kind of physiotherapy, how long is the treatment applied, and how often in the last 4 weeks ....

Figure 7.1 Referral form.

In an ideal world, the referral form is available at the time, when the appointment for the infrared investigation is made. This provides the information needed to select the adequate room temperature for the investigation, and perform necessary quality assurance tests on the equipment.

7.2 Starting the Equipment

At minimum, 30 min before the patient arrives at the examination room, the equipment should be switched on. At first the routine standard procedure of quality assurance must be performed. A logbook that documents the time of last calibration of the thermal imager at the national standard laboratory must be available on site of the examination room. A schedule of minor standard tests for assessing the performance of the equipment must be part of the logbook. Scheduled test must be conducted on time and the test results must be recorded accordingly. In Europe, yearly recalibration of medical temperature-measuring devices is required due to the European directive on medical devices. However, in special conditions such in the case of monitoring blood flow with a thermal imager, the device will not be regarded as a temperature-measuring device [1].

An external blackbody reference source is highly recommended to control for measurement performance of the thermal imager, although such a device might not be always available. Calibration mechanisms within the infrared camera cannot fully compensate a missing external temperature reference. Plassmann has proposed a simple test to assess the offset variation over temperature measurement range by using a container with water that is stepwise cooled down from 40°C to 20°C, while the temperature readings by the camera recorded with perpendicular view on the water surface are compared with the temperature results of a platinum resistance (or a precise mercury) thermometer. This and the other tests proposed should be performed at a time interval of 6 months [2].

Like a thermal imager, an external temperature reference source is not immediately at stable temperature. Depending on the specifications and design of the reference source, a time period between 15 and 45 min may be needed for stability. In a busy examination centre, it might be useful to have the reference source permanently switched on.

7.3 Room Temperature

The room temperature must be displayed in the examination room. When the target temperature is reached, its variation must be recorded over the time of the performed imaging session. Variation of room temperature should be within 1°C.

7.3.1 Choice of Ambient Temperature

As we already know, the temperature at the surface a human body is not evenly distributed caused by physiological and/or pathological processes. The underlying condition may result in either in local increase (hyperthermia) or in a local decrease of temperature (hypothermia).

A room temperature of 20±0.5°C may be appropriate for expected local hyperthermia as the heat preserving reaction of partly uncovered skin to this slightly cool environment will increase the contrast between unaffected skin with slightly constricted vessels and the dilated vasculature of the affected tissue. Vascular reactions to thermal stimuli seem to be hampered in conditions presenting with maximal or submaximal dilated or constricted vessels.

A room temperature of 24±0.5°C may be appropriate for expected local hypothermia as no heat dissipating or preserving reaction of partly uncovered skin will occur in this almost indifferent thermal environment. However, in the examination of totally undressed, naked subjects, the room temperature should be 22±0.5°C for suspected inflammatory conditions and 26±0.5°C for the identification of reactive temperature changes.

7.3.1.1 Accommodation time

A minimum time of 15 min is required for room temperatures between 20°C and 24°C, and 20 min for a room temperature of 26°C.

7.4 Preparation of the Imaging Equipment for Image Capture

Start the software that controls your camera.

Select recording of single images or series of images at a defined time interval.

Create a data file for the person who should be investigated. Use the information of the referral form for that purpose. Add conditions of image recording (date, time, room temperature, views, use of provocation tests such as cold challenge or provocative body positions). Assign a unique label to each personal data file and to all thermal images related to data file. These are needed to retrieve images for later (off-line) evaluation.

7.5 Patient Preparation

Explain the person to be imaged the reason why thermal imaging is performed and the procedures necessary for recording valid and reliable thermal images.

Read the referral form and ask the person to unclothe the body around the area of interest and other body parts which may functionally related to the body region of the actual complaint. The fields of body views defined in the Glamorgan protocol are described in Chapter 6.

7.5.1 Patient's Position during Acclimation and Image Capture

Different clinical questions require different body positions of the patient during acclimation and the body parts being imaged also differ in various indications for thermal imaging.

7.5.1.1 Indication: Headache including sinusitis

The patient may stand, or sit on a stool, the upper body should be undressed, and glasses and jewellery should be removed. The patient should wear a disposable hat to keep hair away from the face and the dorsal neck. Touching the face and the dorsal neck must be avoided.

Use views 4 (FA) and 5 (ND) for imaging. View 7 (UB) is optional.

7.5.1.2 Indication: Shoulder pain including arthritis, osteoarthritis and rotator cuff derangements

Person may stand, or sit on stool (not on a chair with backrest), the upper body should be undressed, jewellery should be removed from the neck region and arms, in females, and at least the upper straps of bras should be removed. Folding hands or crossing arms must be avoided.

Use views 7 (UB), 8 (RAA), 9 (RAD), 10 (LAA), and 11 (LAD). View 6 (CA) is optional.

7.5.1.3 Indication: Elbow pain including arthritis, osteoarthritis and epicondylitis

Person may stand, or sit on stool (not on a chair with backrest), the total arms should be undressed, and jewellery should be removed from the arms. Folding hands or crossing arms must be avoided.

Use views 7 (UB), 8 (RAA), 9 (RAD), 10 (LAA), and 11 (LAD).

7.5.1.4 Indication: Wrist pain and pain in finger joints including arthritis, osteoarthritis and complex regional pain syndrome (CRPS)

The patient may stand, or sit on stool (not on a chair with backrest), the forearms should be undressed, jewellery should be removed from the forearms and hands, the hands should hang freely on the side of the body, and contact with the lower body and finger movements must be avoided.

Use view 12 (BHD).

7.5.1.5 Indication: Paraesthesia of the upper extremity including carpal tunnel syndrome and thoracic outlet syndrome

Person may stand or sit on stool (not on a chair with backrest), the upper body should be undressed, and jewellery should be removed from the neck region and arms. The hands should hang on the side of the body, and contact with the lower body and finger movements must be avoided.

Use the Thoracic Outlet protocol plus view 12 (BHD).

7.5.1.5.1 *Positions for image capture during TOS-testing*

7.5.1.5.1.1 *Position 1: At rest two alternative positions are possible*

1A. The patient is standing, the arm is slightly abducted (20°), and the palm pointing towards the camera. Start with the right hand side and record the image of the left hand side next.

The camera must be brought to the level of the centre of the palm and must look perpendicularly to the hand (Fig. 7.2).

1B. The patient is sitting in front of table and both hands, palms upwards, are placed on a table. Full supination of forearms may be difficult to achieve in some patients with arthritis, and in such subjects, the thumb may not have full contact with the desk. The camera is placed above the hands, looking perpendicularly to the palm (Fig. 7.3).

Both positions of the hands are below the heart level.

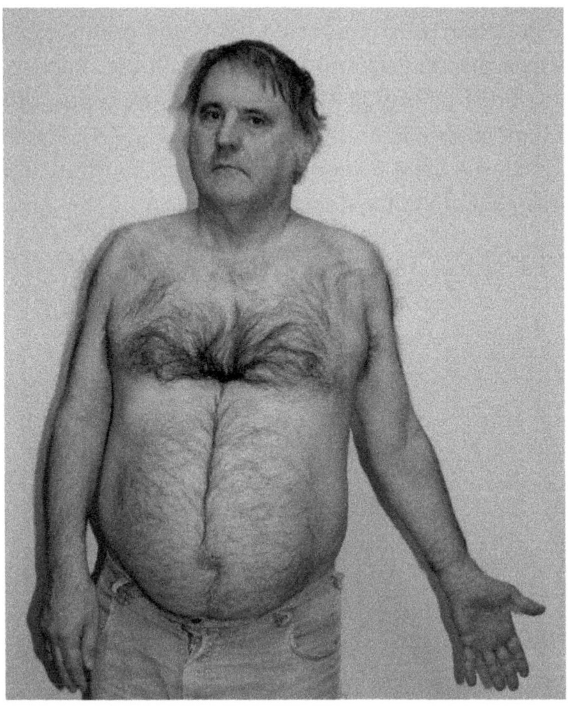

Figure 7.2 Position 1 A for TOS-testing.

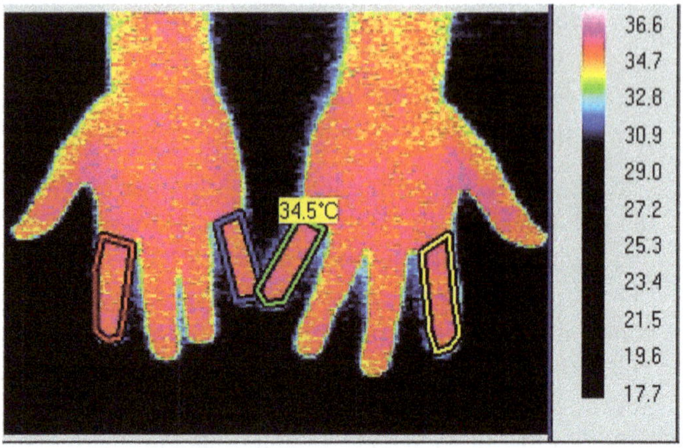

Figure 7.3 Baseline position 1B for TOS-testing.

7.5.1.5.1.2 *Position 2: Hands up*

The patient is standing. Start image recording at the right hand side first. The shoulder is 90° abducted, the hand points upwards, the palm towards the camera and the head is in the normal position. The camera must be brought to the level of the centre of the palm and must look perpendicularly to the hand (Fig. 7.4). Prior to image capture, check the hand position with respect to tilting of the hand and/or head that might have occurred.

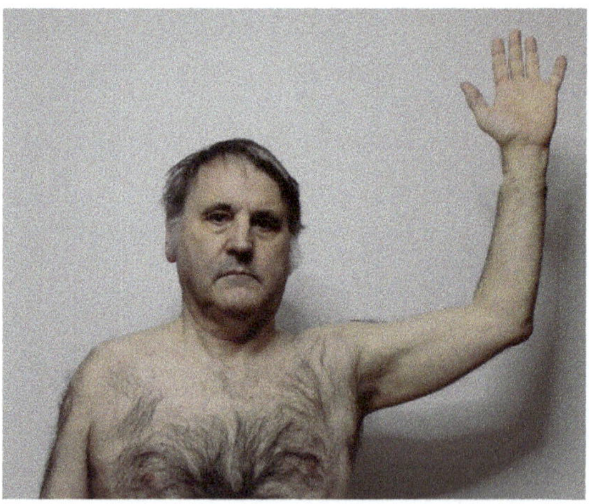

Figure 7.4 Position 2: hands up for TOS-testing.

7.5.1.5.1.3 *Position 3: Hyperabduction test*

Start image recording at the right hand side first. The patient is standing with maximum abduction in the shoulder joint, the elbow flexed 90° with the forearm pointing upwards, head turned the affected side, full inspiration and breath holding for 30 s (Fig. 7.5). The camera must be brought to the level of the centre of the palm and must look perpendicularly to the hand. Prior to image capture, check the arm and hand position with respect to lowering the arm in level and/or tilting of the hand that might have occurred.

Record the image at the immediate end of the breath-holding period having resumed normal breathing.

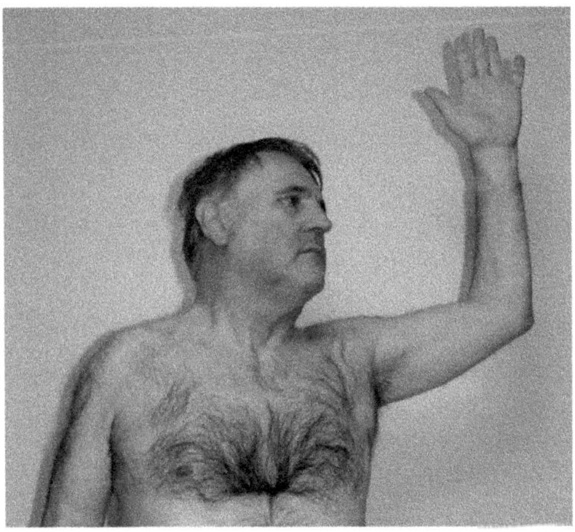

Figure 7.5 Position 3: hyperabduction test for TOS-testing.

7.5.1.5.1.4 *Position 4: Modified Adson's manoeuvre*

Start image recording at the right hand side first. The patient is standing, the arm is slightly abducted at the shoulder (20°), The palm is pointing forward. In maximum inspiration and breath holding for 30 s, the head is turned away from the investigated side (Fig. 7.6). The camera must be brought to the level of the centre of the palm and must look perpendicularly to the hand. Prior to image capture, check the hand position with respect to the abduction angle and/or tilting of the hand that might have occurred. Record the image at the immediate end of the breath-holding period having resumed normal breathing.

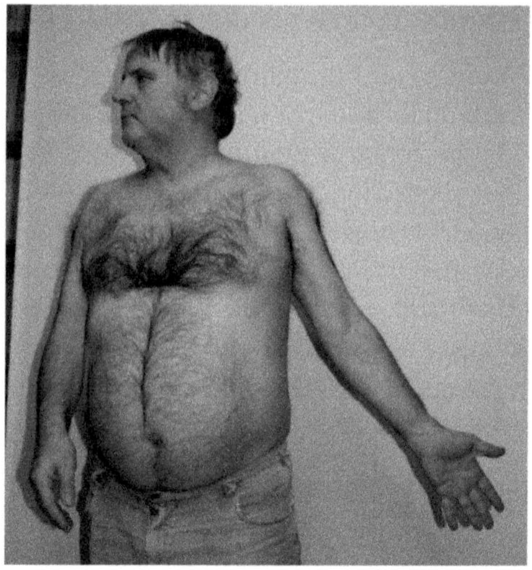

Figure 7.6 Position 4: modified Adson's manoeuvre for TOS-Testing.

7.5.1.6 Indication: Raynaud's phenomenon including vibration induced white fingers

The thermographic examination in patients with suspected Raynaud's phenomenon requires the application of a mild cold challenge.

7.5.1.6.1 *Preparation of the cold challenge*

7.5.1.6.1.1 *Choose the water temperature*

For primary and secondary Raynaud cases, 20°C is an accepted temperature. For testing vibration-induced fingers, 15°C is an established standard.

7.5.1.6.1.2 *Prepare the cold water bath*

Start to fill a basin or a bucket with water of 20°C (or 15°C used by a few authors). The size of the water container should provide the immersion of hands up to the wrist. Most convenient is water basin with a drain and two taps supplying warm and cold water. Use an immersion thermometer for checking the water temperature. Usually the cold tap water gets colder the longer it is supplied. Fill the basin with cold water. Read the water temperature after stirring the water with your thermometer to avoid non-accurate temperature readings. If the water is too warm, remove a portion of the water by opening the drain plug while continuously adding cold water. Close the drain and check the temperature again. Repeat the procedure until the water temperature is equal to 20°C. If the temperature is too low, add water from the warm water tap.

7.5.1.6.2 *Patient's position during acclimatization and image capture*

The patient may stand, or sit on stool (not on a chair with backrest), the forearms should be undressed, jewellery should be removed from the forearms and hands, the hands should hang freely on the side of the body, and contact with the lower body and finger movements must be avoided.

After acclimation, both hands with palms downwards are placed on a table and an image in view 12 (BHD). The camera is placed above the hands, looking perpendicularly to the dorsal surface of the hands.

7.5.1.6.3 *Cold challenge*

After the first image has been recorded, the patient puts on plastic gloves and immerses the hands up to the wrists into the water bath for 1 min (or 3 min after 15°C cooling). After the immersion period, the gloves must be removed by the technician. If wetness of the skin occurs (in case of glove leakage), the technician should gently dry the skin with a disposable towel.

Bring the hands of the patient in a similar position as in the baseline image. Explain to the patient to keep the hands without moving the fingers in the same place during the total imaging session. Record a series of five images at a time interval of 5 min, with the first image taken immediately after repositioning following the cold challenge.

7.5.1.7 Indication: Low back pain, including iliosacral arthritis

The patient is standing (not sitting), and lower back, abdomen gluteal region and thighs should be undressed. If pants are worn, they should be minimized to smallest possible size. Contacts with the environment in the lower body region and leg movements must be avoided.

Use views 13 (ABD), 14(LB), 16 (TD), 17 (LRL) and 18(LLL). View 15(TA) is optional.

7.5.1.8 Indication: Sciatica including detection of varicose veins of the leg

The patient is standing on a heat insulating floor mat, lower back, abdomen and legs should be undressed. If pants are worn, they should be minimized to smallest possible size. Contacts with the environment in the lower body region and leg movements must be avoided.

Use views 15 (TA), 16 (TD), 17 (LRL), 18 (LLL), 19 (LLA), 20 (LLD) 23(DF) and 24(DF). The lower part of total body views 1 (TBA) and 2 (TBD) are optional. Lower part is equal to the image of the total body below the waist.

7.5.1.9 Indication: hip or groin pain

The patient is standing (not sitting), lower back, abdomen, gluteal region and thighs should be undressed, and pants should be removed. Leg movements and contacts with the environment in the lower body region must be avoided.

Use views 13 (ABD), 16 (TD), 15(TA) and 16 (TD). Views 17 (LRL) and 18 (LLL9 are optional.

7.5.1.10 Indication: Knee pain including arthritis and osteoarthritis of the knee, complex regional pain syndrome (CRPS) and soft tissue trauma

The patient is standing (not sitting), and legs must be undressed. Contacts of the legs with the environment and leg movements must be avoided.

Use views 21 (BKA), 17 (LRL) and 18(LLL). Views 15(TA), 16 (TD), 18 (LLA) and19 (LLD) are optional.

7.5.1.11 Indication: Ankle pain including arthritis, osteoarthritis of the ankle, complex regional pain syndrome (CRPS) and soft tissue trauma

The patient may stand or sit, feet are in contact with a heat insulating floor mat, lower legs must be undressed. Contacts of the legs with the environment and leg movements must be avoided.

Use views 22 (BAA) and 23 (DF). Views 18 (LLA) and19 (LLA) are optional.

7.5.1.12 Indication: Foot pain including arthritis, osteoarthritis of the midfoot, and toes, complex regional pain syndrome (CRPS), peripheral neuropathy

The patient may stand or sit on a chair, and the feet are placed on a heat insulating floor mat. When sitting, the lower leg might be supported by a foot stool avoiding in this way any contact to the sole. Lower legs must be undressed. Contacts of the lower legs with the environment and leg movements must be avoided.

Use views 23 (DF) and 24 (PF); views 18 (LLA) and19 (LLD) are optional.

7.5.1.13 Indication: Posterior foot pain including pain related to the Achilles tendon

The patient may sit on a chair, and the feet are placed a heat insulating floor mat; lower legs must be uncovered. Contacts of the lower legs with the environment and leg movements must be avoided.

Use views 19 (LLD) and 24 (PF).

7.6 Image Capture

Before you record the image, check whether the camera looks perpendicularly to the body region being imaged, the subject is in the proposed position for recording this particular body region, and the field of views meets the required features such as anatomical landmarks and boundaries when extremities are imaged. Always record an image of the right and the left body side.

Press the respective key to record the image and label the file accordingly.

7.7 Image Evaluation

Retrieve the recorded images, choose measuring areas from the catalogue of 96 regions of interest (ROI) presented in Chapter

6. Generate size, shape and location of the ROI accordingly to the description provided in the catalogue. Regions of interest must be placed on both sides of corresponding body regions.

Record temperature values provided for the regions of interest: Minimal request is mean temperature±standard deviation; maximum and minimum temperatures are optional. The difference in the mean temperature of corresponding regions of interest must be calculated.

7.7.1 Evaluation of Thermal Images from Patients with Suspected Thoracic Outlet Syndrome

In patients with suspected thoracic outlet syndrome, the regions of interest are placed on the index and the little finger. The outline of the measuring shape crosses the index (little finger) at the inter-digital fold, follows the outline of the finger and returns to the starting point. Such ROIs must be generated for each image of thoracic outlet series (Position $1a_1$: right palm $1a_2$: left palm or Position 1b: both palms placed on a table, Position 2: hands up. 2_1 right hand, 2_2: left hand, Position 3: hyperabduction, 3_1: right hand, 3_2: left hand, Position 4: Modified Adson test, 4_1: right hand, 4_2: left hand). Figure 7.7 shows a TOS series.

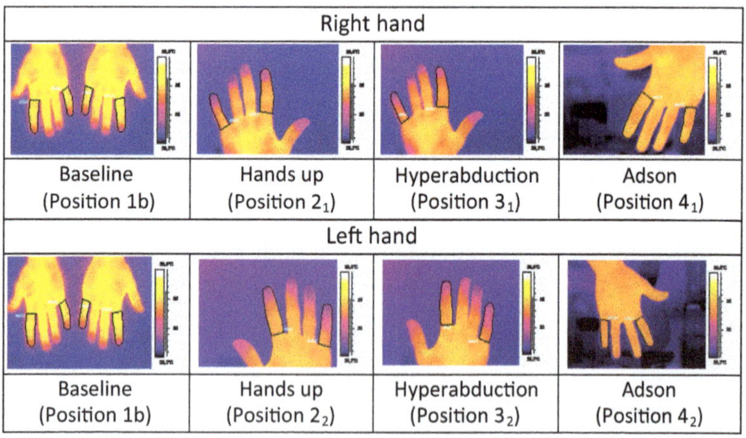

Figure 7.7 Typical findings of thermographic investigation in suspected TOS.

The difference in the mean temperature of the index finger and the little finger is less than 0.5°C in healthy subjects; differences of 0.5°C or more are regarded as abnormal. Based on such temperature differences, the following classification was made:

An abnormal temperature difference in one position was named "possible TOS", abnormal findings in two positions were regarded as a "probable TOS", and abnormal temperature readings in three or four arm position were classified as "definite TOS". In addition, TOS cases were labelled with "index" or "little finger" with respect to the colder finger. Table 7.1 shows the evaluation of the image series of Fig. 7.7.

Table 7.1 Temperature readings from the thermal images of Fig. 7.7

	Baseline (°C)	**Hands up (°C)**	**Hyperabduction (°C)**	**Adson (°C)**
		Right hand		
Index finger	34.4	32.5	32.4	33.6
Little finger	34.1	31.5	31.3	33.2
Temperature difference	0.31	0.99	-1.72	0.37
		Left hand		
Index finger	34.4	32.3	32.5	33.6
Little finger	34.2	31.1	31.2	33.2
Temperature difference	0.24	1.22	1.24	0.35

7.7.2 Evaluation of Thermal Images from Patients with Suspected Carpal Tunnel Syndrome

In cases of suspected carpal tunnel syndrome, the same regions of interest (index and little finger) should be used and applied to the thermal image of the dorsal hands.

7.7.3 Evaluation of Thermal Images from Patients with Suspected Raynaud's Phenomenon

In patients with suspected Raynaud's phenomenon, the temperature of each fingertip and the temperature above the middle of each metacarpal bone must be determined.

7.7.3.1 Position of ROIs at the fingertip

Circularly shaped regions of interest are used for that purpose, the diameter of these circles is the same as the width of each finger pad, and the circle is aligned to the outline of the fingertip.

7.7.3.2 Position of ROIs at the metacarpal bones

Use auxiliary lines to find the correct position of the circular Regions of interest. Use the same diameter as for the circles at the fingertip.

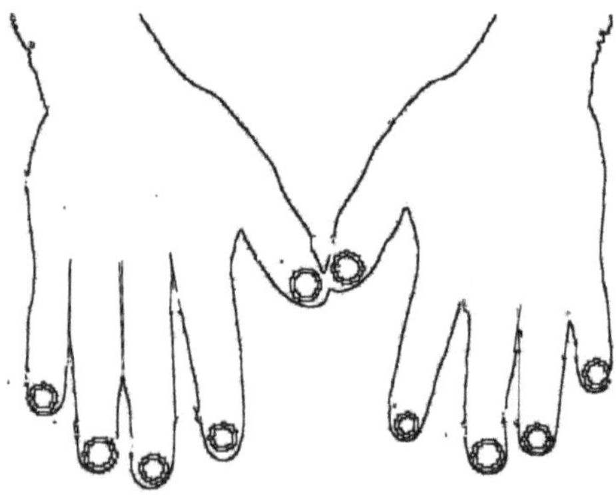

7.7.3.2.1 *Auxiliary lines*

A system of auxiliary lines supports a reliable positioning of ROIs at the dorsum of the hand.

7.7.3.2.1.1 *Step 1*

Draw line from the middle of the outline of fingertip to the wrist joint following the median axis of each finger. This auxiliary line is equal to the X-Section of individual fingers as described in Chapter 6.

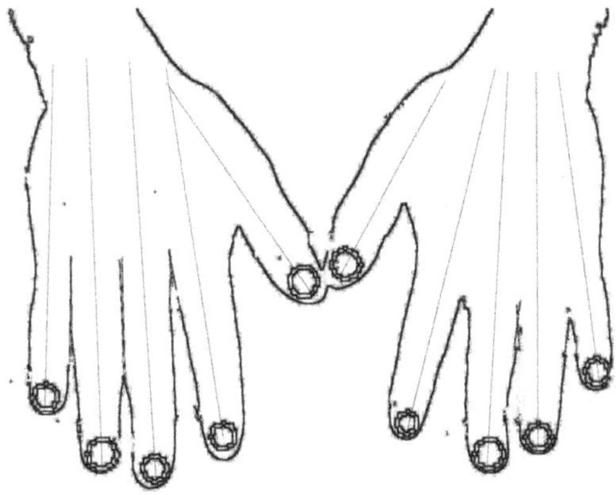

7.7.3.2.1.2 *Step 2*

When the fingers are spread, the basic phalanx of the index finger and the second metacarpal bone form an open angel on the radial side, while the basic phalanx of the little finger and the fifth metacarpal bone form an open angel on the ulnar side. Connect the apex of these angles with a line.

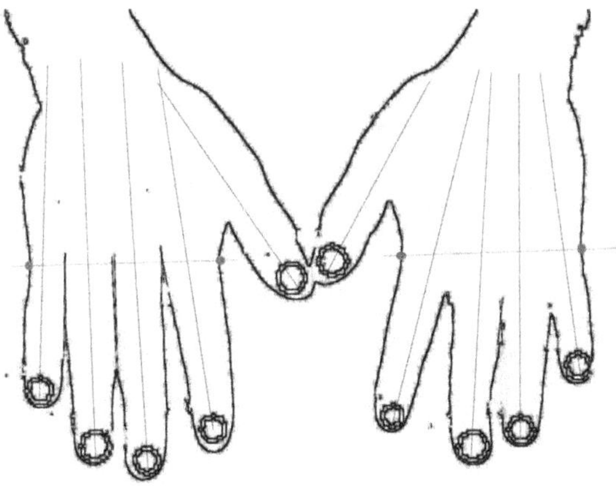

7.7.3.2.1.3 *Step 3*

Also, the forearm and the carpal bones form open angels at the radial and the ulnar side of the wrist. Connect these angular points with a line.

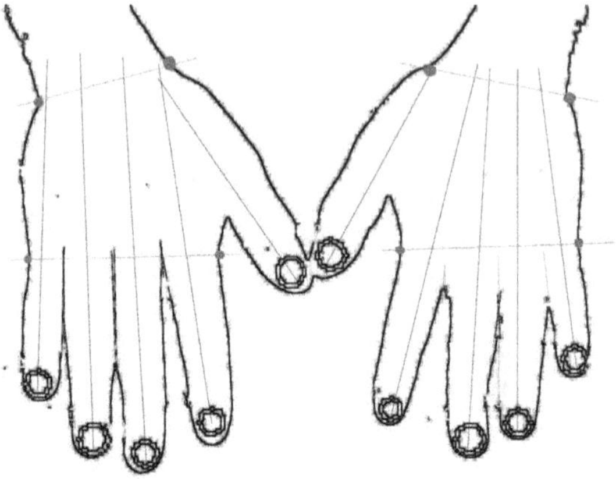

7.7.3.2.1.4 *Step 4*

Find the first metacarpal joint. Draw a perpendicular line from the centre of the circumference of first metacarpal joint to the axis of the thumb.

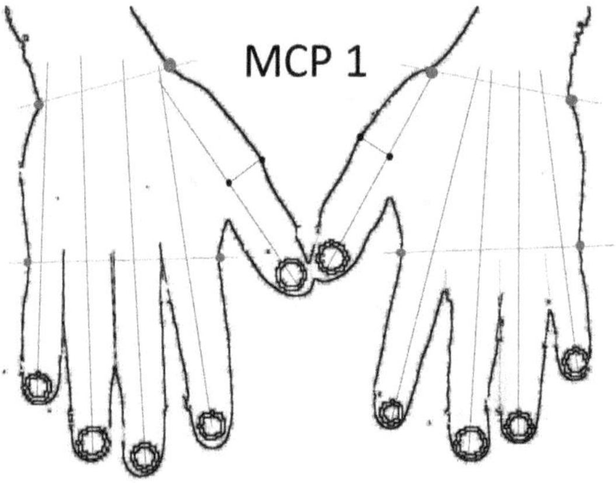

7.7.3.2.1.5 *Step 5*

Find the section of each median finger axis extending between the auxiliary lines at the metacarpal joints and the wrist. Divide this distance into equal parts.

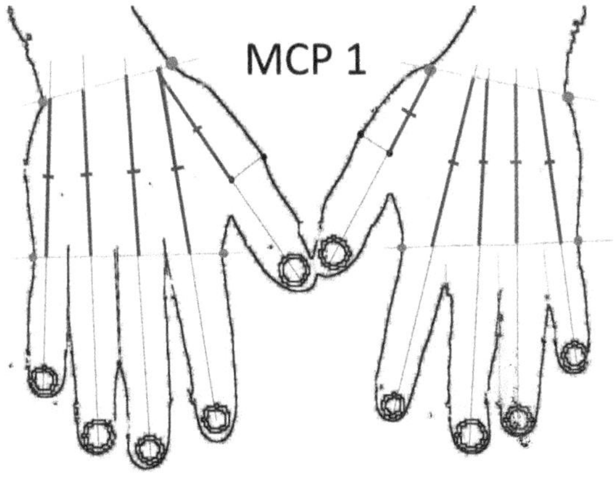

7.7.3.2.1.6 *Step 6*

Align the centre of the metacarpal ROIs to middle of the distance between the two auxiliary lines of the finger axis.

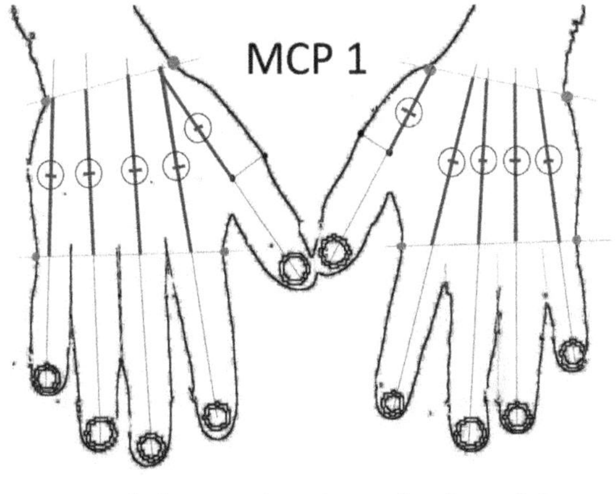

It is recommended to novices in evaluation of thermal images recorded from Raynaud patients to use the proposed auxiliary lines regularly for achieving reproducible temperature readings.

7.7.4 Calculation of Temperature Gradients

A longitudinal temperature gradient of each finger and a combined temperature gradient, summarizing the temperature gradients of individual fingers from different time points.

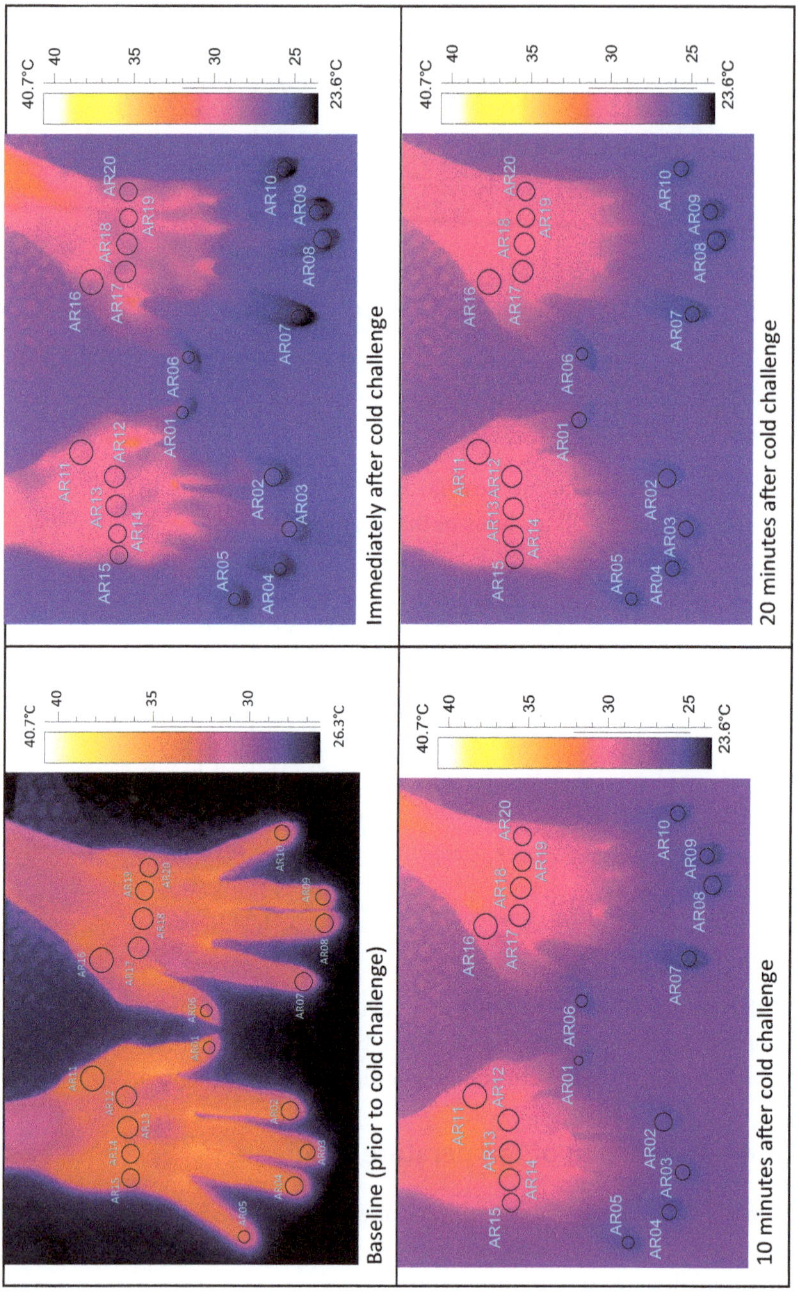
Baseline (prior to cold challenge)
40.7°C
40
35
30
26.3°C
Immediately after cold challenge
40.7°C
40
35
30
25
23.6°C
10 minutes after cold challenge
40.7°C
40
35
30
25
23.6°C
20 minutes after cold challenge
40.7°C
40
35
30
25
23.6°C
AR01
AR02
AR03
AR04
AR05
AR06
AR07
AR08
AR09
AR10
AR11
AR12
AR13
AR14
AR15
AR16
AR17
AR18
AR19
AR20

Baseline (prior to cold challenge)

finger/site	1.right	2.right	3.right	4.right	5.right	1.left	2.left	3.left	4.left	5.left
fingertip	32.6	33.0	33.3	33.2	32.7	32.3	31.8	32.7	32.5	32.5
metacarpal	33.2	32.9	33.3	33.6	33.3	32.6	32.6	32.9	33.2	32.9
gradient	–0.68	0.04	–0.02	–0.37	–0.58	–0.34	–0.72	–0.17	–0.69	–0.38

10 min after cold challenge

finger/site	1.right	2.right	3.right	4.right	5.right	1.left	2.left	3.left	4.left	5.left
fingertip	26.4	25.9	25.9	25.4	25.4	25.9	25.1	25.3	25.1	25.3
metacarpal	30.7	30.8	31.0	30.9	29.8	30.3	30.4	31.0	30.8	30.1
gradient	–4.31	–4.93	–5.18	–5.55	–4.34	–4.34	–5.32	–5.74	–5.68	–4.76

20 min after cold challenge

finger/site	1.right	2.right	3.right	4.right	5.right	1.left	2.left	3.left	4.left	5.left
fingertip	26.3	25.7	25.5	25.2	25.2	25.7	24.9	25.0	24.9	25.2
metacarpal	30.0	30.1	30.3	30.3	29.3	29.8	29.9	30.3	30.3	29.7
gradient	–3.73	–4.44	–4.84	–5.14	–4.07	–4.13	–4.99	–5.37	–5.32	–4.52

Combined gradients

gradient/time	1.right	2.right	3.right	4.right	5.right	1.left	2.left	3.left	4.left	5.left
prior + 10 min	–4.99	–4.89	–5.2	–5.92	–4.92	–4.68	–6.04	–5.91	–6.37	–5.14
prior + 20 min	–4.41	–4.4	–4.86	–5.51	–4.65	–4.47	–5.71	–5.54	–6.01	–4.9

Figure 7.8 Typical findings in a case of suspected Raynaud's phenomenon.

7.7.4.1 Temperature gradient metacarpal bone to fingertip

At each finger, subtract the mean temperature of metacarpal bone from the mean temperature of the fingertip. In case of cold fingertips, the resulting gradient is negative, but in case of warm fingertips, the gradient gets positive.

7.7.4.2 Combined temperature gradient

Calculate the combined gradients by summarizing the values of gradients prior to the cold challenge with the gradients 10 min or 20 min after the cold challenge. This results in a total of 20 temperature gradients, 2 for each finger.

All images, all temperature readings and all gradient calculations must appear in the report. The result section of such a report may look like the one shown in Fig. 7.8.

7.8 Edit a Report

Finally, the findings obtained with infrared thermal imaging must be communicated to the referring physician as a written report. Such a report must include the following required information.

7.8.1 Name of the Referring Physician

The name of the health professional who requested a clinical thermal image must be included in the report.

7.8.2 Personal Information of the Patient Including the Health Problem That Led to the Referral

All this information is available from the referral form, which may be copied and pasted building in that way the first section of the report.

7.8.3 Imaging Conditions

All environmental conditions such as brand of the thermal imager, room temperature, time for acclimation, patient's position and clothing during the acclimation period must be reported. Any provocation test such as exposure to a cold challenge or particular positions of imaged body parts must be described in detail.

7.8.4 Captured Images

All images recorded must be displayed in the report. As mentioned in Section 7.4, all images should be labelled with date, room temperature, abbreviation of body view and/or provocation test. All images must be presented together with the applied regions of interest (Fig. 7.8).

If you compare your actual temperature readings with temperature values taken from reference populations, check a possible bias in that comparison and quote your reference.

7.8.5 Temperature Readings

Minimal request is to report mean temperature±standard deviation of each region of interest, maximum and minimum temperatures are optional. The difference in mean temperature of corresponding regions of interest must be calculated. Presentation of temperature value in a table is recommended (Fig. 7.9).

7.8.6 Interpretation and Recommendations for Further Investigations

In most national health systems, only medical physicians have the privilege to draw **diagnostic conclusions** from thermal images and recommend further investigation based on the thermal findings.

When thermal imaging is performed to assess health conditions, for example, muscle function or cardiovascular endurance, the respective health professional may interpret skin temperature in relationship to body function of interest but must not establish any relationship to diseases.

7.8.7 Approve and Sign Your Report

Finally, check your report for spelling errors and more important, if any required information is missing. Add date of editing and sign the report.

Patient: N. I, female, age. 73 years
Referring Physician: Dr Ammer
Reason for Referral: CRPS at the right hand?

After acclimatization to a room temperature of 21°C for 15 min with bare forearms, static thermal images of both hands were recorded, and the following temperatures were obtained.

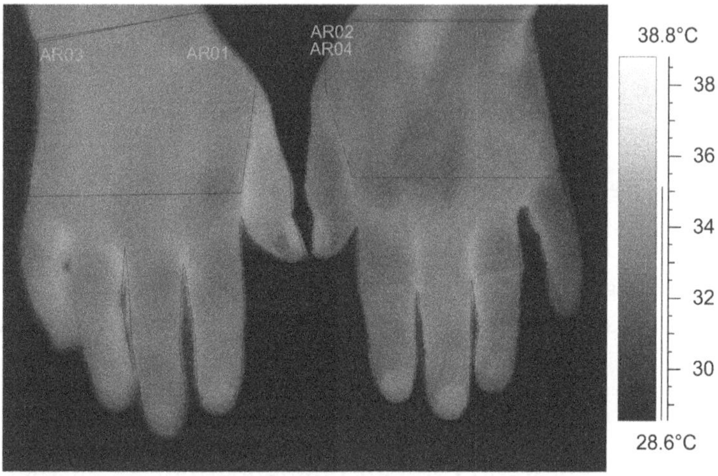

The right hand appeared to be warmer than the left hand, the temperature difference of the total hand (AR01 –AR02) is 0.75°C, of the dorsal hand (AR03 –AR04) it is 1.02

Measurement area		Temperature [°C]	Standard deviation
AR01	right hand, total	33.3	0.7
AR02	left hand, total	32.5	0.8
AR03	right dorsal hand	33.5	0.4
AR04	left dorsal hand	32.5	0.5
fo01	(ar1.avg) – (ar2.avg)	0.75	
fo02	(ar3.avg) – (ar4.avg)	1.02	

Conclusion
Hyperthermia of the right hand, the temperature difference found at the dorsal hands meets the temperature criteria [1] for the diagnosis of CRPS.

Reference
1. Wilson PR, Low PA, Bedder MD, Covington EC, RaucK RL. Diagnostic Algorithm for Complex Regional Pain Syndromes. In: Jänig W, Stanton-Hicks M, eds, *Reflex Sympathetic Dystrophy: A Reappraisal. Progress in Pain Research and Management*. Vol. 6, IASP Press, Seattle, 1996, pp. 93–105.

Figure 7.9 Example of a thermographic report.

References

1. European Commission, DG Health and Consumer, Directorate B, Unit B2 "Cosmetics and medical devices" Medical Devices: Guidance document- Classification of medical devices. MEDDEV 2,4/ Rev 9, June 2010.
2. Plassmann P, Ring EFJ, Jones CD. Quality assurance of thermal imaging systems in medicine. *Thermology International* 2006, 16(1), 10–15.
3. Ammer K. The sensitivity of infrared imaging for diagnosing Raynaud's phenomenon and for Thoracic Outlet Syndrome is dependent on the method of temperature extraction from thermal images. *Thermology International* 2008, 18(3), 81–88.
4. Ammer K. The Glamorgan Protocol for recording and evaluation of thermal images of the human body. *Thermology International* 2008, 18, 125–144.
5. Ammer K, Ring F. Standard procedures for infrared imaging in medicine. In: Diakides NA, Bronzino JB, eds, *Medical Infrared Imaging*, CRC Press, 2008, pp. 22.1–22.14.

Index

For Product Safety Concerns and Information please contact our EU representative GPSR@taylorandfrancis.com
Taylor & Francis Verlag GmbH, Kaufingerstraße 24, 80331 München, Germany

www.ingramcontent.com/pod-product-compliance
Ingram Content Group UK Ltd.
Pitfield, Milton Keynes, MK11 3LW, UK
UKHW022159190926
475702UK00003B/99
* 9 7 8 9 8 1 4 7 4 5 8 2 6 *